Chemical Calculations

Uniquely organized by chemical rather than mathematical topics, this book relates each mathematical technique to the chemical concepts where it applies. The new edition features additional, revised, and updated material in every chapter and maintains the clarity of the previous edition with the appropriate organization of topics and improved cross-referencing where mathematical techniques occur more than once. The text contains additional worked examples and end-of-chapter exercises with detailed solutions—giving students the opportunity to apply previously introduced techniques to chemically related problems. It is an ideal course companion for chemistry courses throughout the length of a degree.

Features

- This book covers the difficult area of mathematics in an easy-to-read format for students and professionals in chemistry and related subjects.
- Structured according to chemical rather than mathematical topics.
- Each topic has at least 12 end of chapter applied chemistry problems to provide practice in applying the techniques to real chemistry.
- Indexing of material by both chemical and mathematical topics.
- Extends its utility as a concise and practical reference for professionals in a wide array of scientific disciplines involving chemistry.

Chemical Calculations
Mathematics for Chemistry

Third Edition

Paul C. Yates
Deputy Director of Learning, Teaching and Scholarship

CRC Press
Taylor & Francis Group
Boca Raton London New York

CRC Press is an imprint of the
Taylor & Francis Group, an **informa** business

First edition published 2023
by CRC Press
6000 Broken Sound Parkway NW, Suite 300, Boca Raton, FL 33487-2742

and by CRC Press
4 Park Square, Milton Park, Abingdon, Oxon, OX14 4RN

Library of Congress Cataloging-in-Publication Data
Names: Yates, Paul, 1961- author.
Title: Chemical calculations : mathematics for chemistry / Paul C. Yates,
Deputy Director of Learning, Teaching and Scholarship.
Description: Third edition. | Boca Raton : CRC Press, 2023. | Includes bibliographical references and index. |
Summary: "Uniquely organized by chemical rather than mathematical topics, this book relates each mathematical technique to the chemical concepts where it applies. The new edition features additional, revised, and updated material in every chapter, and maintains the clarity of the previous edition with appropriate organization of topics and improved cross-referencing where mathematical techniques occur more than once. The text contains additional worked examples and end-of-chapter exercises with detailed solutions-giving students the opportunity to apply previously introduced techniques to chemically related problems. An ideal course companion for chemistry courses throughout the length of a degree. This book covers the difficult area of mathematics in an easy-to-read format for students and professionals in chemistry and related subjects. Structured according to chemical rather than mathematical topics. Each topic has 10 end of chapter applied chemistry problems to provide practice in applying the techniques to real chemistry. Indexing of material by both chemical and mathematical topics. Extends its utility as a concise and practical reference for professionals in a wide array of scientific disciplines involving chemistry"— Provided by publisher.
Identifiers: LCCN 2022030163 (print) | LCCN 2022030164 (ebook) | ISBN 9780367488673 (hardback) | ISBN 9780367488666 (paperback) | ISBN 9781003043218 (ebook)
Subjects: LCSH: Chemistry—Mathematics.
Classification: LCC QD39.3.M3 Y38 2023 (print) | LCC QD39.3.M3 (ebook) | DDC 540.1/51—dc23/eng20221031
LC record available at https://lccn.loc.gov/2022030163
LC ebook record available at https://lccn.loc.gov/2022030164

ISBN: 978-0-367-48867-3 (HB)
ISBN: 978-0-367-48866-6 (PB)
ISBN: 978-1-003-04321-8 (EB)

DOI: 10.1201/9781003043218

Typeset in Warnock Pro
by codeMantra

Dedication

Paul Yates wants to dedicate this book to Julie, Catherine and Christopher

Contents

Preface

I was delighted to be invited to produce a third edition of this book, not least because that suggests that students have found the previous editions of use. Since the second edition was published, I have had the privilege of working on national projects in the area of mathematics support for chemists and others, and this has convinced me that the need for textbooks such as this is as great as ever.

Those who were familiar with the previous edition will see that its structure and underlying philosophy remain the same. It is structured as it would be a textbook in physical chemistry, and this drives the presentation of mathematics. This ensures that all the mathematics is relevant to chemistry, and I have strengthened this link in a number of places.

Changes are relatively minor; I have attempted to eliminate any errors found in the previous edition, although I appreciate that this goal is like a mathematical limit, in that it can never be fully attained. I apologise in advance for those that remain and hope that you will still find the book useful. Wording has been changed in places where I didn't feel that the text was quite as clear or as friendly as it might be.

I have included more links between sections, which I think is particularly important as similar mathematical topics may be physically separated by virtue of the chemistry that they support. There are some new topics, but the main additions have come in the form of additional exercises and problems. I believe that a book in this area will only achieve its aim of developing mathematical skills if its readers are given sufficient opportunity to practise those skills. As in previous editions, exercises test the raw mathematics while the problems are more chemically related.

Finally, although you may not find mathematics the easiest subject to study or indeed have expected to need to do so when you first decided to study the chemical sciences, I hope you will persevere. Mastery of the topics in this book will open up a whole range of chemistry and give you insights across the subject, not just in physical chemistry. I wish you well with your studies.

Paul Yates
Staffordshire, United Kingdom
June 2022

Acknowledgement

Since the publication of the second edition, I have had the pleasure of working in a national role which has included involvement in a number of projects and workshops for the support of mathematics both within chemistry and across the physical sciences and STEM subjects. I would like to thank all the colleagues and students who have influenced my thinking about the subject during that time. Some of those colleagues have reviewed the proposal for this edition, and their input has helped to shape both its scope and content. I would also like to thank my current colleagues for many interesting discussions as to how the teaching of mathematics, both in and beyond the context of chemistry, compares with that in the arts and humanities and social sciences.

I would like to thank Hilary Lafoe and her colleagues at Taylor and Francis for their belief that another edition was appropriate and their support throughout its production.

Above all, I would like to thank my family for their continued love, support, and understanding.

Paul Yates
Staffordshire, United Kingdom
June 2022

Author

Paul Yates holds a BSc in Chemical Physics, a PhD in Chemistry and an MA in Learning and Teaching in Higher Education. After several years lecturing in Physical Chemistry at Keele University, he moved into educational development. He was subsequently able to combine this experience in the post of Discipline Lead for the Physical Sciences at the Higher Education Academy.

He has a long-standing interest in the development of mathematical skills and is the author of two textbooks on mathematics for chemists. Since returning to the university sector, he has developed an interest in the way in which data and metrics are used by various stakeholders including student supporters.

He received a Keele University Excellence in Teaching Award. He is a Fellow of the Royal Society of Chemistry, a Senior Fellow of the Staff and Educational Development Association, and a Principal Fellow of the Higher Education Academy.

Fundamentals

1.1 INTRODUCTION

There are some mathematical processes without which you will find it very difficult to solve any problems, no matter how good your grasp of the other concepts involved. These are placed in this preliminary chapter so that you can find them easily if you need to refer to them while studying other parts of this book. Even if you are familiar with this material you may find it useful to see how these concepts are used in real chemical examples.

1.2 POSITIVE AND NEGATIVE NUMBERS

If a number is represented by the algebraic symbol a, its negative will be written as $-a$. For example, the negative of 4 is -4. There are two rules to remember when combining numbers which may be positive or negative: two like symbols combine to give the positive, whereas two unlike symbols combine to give the negative. This leads to the following sets of rules.

1.2.1 ADDITION

If you are adding a negative number to any other number, this is the same as subtracting the corresponding positive value. For example

$$2 + (-3) = 2 - 3 = -1$$

since the $+$ and $-$ signs are different and therefore give - for the answer. If the two numbers are represented algebraically as a and b then the rule can be written as

$$a + (-b) = a - b$$

This simple rule finds application in chemistry when we wish to determine the overall change in entropy for a process. Entropy is denoted by the symbol S, while a change in entropy is denoted as ΔS. The latter makes use of the symbol Δ to represent the Greek letter delta.

DOI: 10.1201/9781003043218-1

We frequently meet this symbol in physical chemistry to denote a change in quantity. The entropy of a substance indicates the extent of disorder.

The overall entropy change is denoted as $\Delta_{universe}S$, where the term universe is taken to represent both the system of interest and its surroundings. The overall entropy change is then given by

$$\Delta_{universe}S = \Delta_{system}S + \Delta_{surroundings}S$$

where $\Delta_{system}S$ and $\Delta_{surroundings}S$ represent the entropy changes of the system and its surroundings, respectively. Notice that here we are using the subscripts universe, system and surroundings to indicate the actual difference that Δ refers to.

When 2.00 mol of supercooled water at −15.0°C is frozen irreversibly at a constant pressure of 1.00 atm to give ice at −15.0°C the entropy changes in the system and the surroundings are 39.61 and 42.03 J K^{-1}, respectively. The overall entropy change of the universe will then be

$$\Delta_{universe}S = \Delta_{system}S + \Delta_{surroundings}S$$

$$= (-39.61 + 42.03) \text{ J K}^{-1}$$

$$= 2.42 \text{ J K}^{-1}$$

It is worth pointing out here that we can only add or subtract quantities which are expressed in the same units.

1.2.2 SUBTRACTION

If you are subtracting a negative number, this is equivalent to adding the corresponding positive number. For example

$$2 - (-4) = 2 + 4 = 6$$

since there are two like signs combining to give +. Using the symbols a and b we can write the general rule as

$$a - (-b) = a + b$$

We make use of this in chemistry when calculating the expected frequency of a line in an electronic spectrum. The difference ΔE in energy between two levels E_1 and E_2 is given by

$$\Delta E = E_2 - E_1$$

and ΔE is proportional to the frequency of the resulting line. Frequency is converted to energy by multiplying with the Planck constant h; the value of this and other physical constants can be found in Appendix B.

The energy levels in an atom are usually measured relative to the zero energy of a free electron, which means that they are negative.

In hydrogen, the lowest energy level $E_1 = -2.182 \times 10^{-18}$ J and $E_2 = -0.546 \times 10^{-18}$ J. In this case

$$\Delta E = E_2 - E_1$$

$$= -0.546 \times 10^{-18} \text{ J} - (-2.182 \times 10^{-18} \text{ J})$$

The adjacent negative signs combine to give a positive, so

$$\Delta E = -0.546 \times 10^{-18} \text{ J} + 2.182 \times 10^{-18} \text{ J}$$

Since the result of an addition doesn't depend on the order of the numbers being added, we can rewrite this as

$$\Delta E = 2.182 \times 10^{-18} \text{ J} - 0.546 \times 10^{-18} \text{ J}$$

$$= 1.636 \times 10^{-18} \text{ J}$$

It is the commutative law which states that $a + b = b + a$, or that a sum is unaltered by changing the order of its constituent terms.

1.2.3 MULTIPLICATION

When two positive numbers a and b are multiplied, the result is another positive number, as in

$$3 \times 4 = 12$$

We can represent this in symbols as

$$a \times b = ab$$

where it is usual to omit the multiplication sign $\times$.

If only one of the numbers is negative, then the result will be negative, as in

$$-3 \times 4 = -12 \text{ and } 3 \times -4 = -12$$

These are expressed in symbols as

$$(-a) \times b = -ab \text{ and } a \times (-b) = -ab$$

Brackets are used here to clarify where a negative sign is associated with a quantity. This is often a useful technique for making complicated expressions clearer without changing their mathematical meaning.

If, on the other hand, both numbers are negative, then we have two like signs and the result is positive, as shown by

$$(-3) \times (-4) = 12$$

The general rule is expressed as

$$(-a) \times (-b) = ab$$

An area of chemistry where multiplication involving negative numbers occurs is in the calculation of the change ΔG in Gibbs free energy for a process. ΔG is zero at equilibrium, negative for a spontaneous process and positive for a non-spontaneous process, as shown in Figure 1.1. $\Delta G^{\ominus}$ is the value of ΔG when all the reacting species are in their standard states, defined as their most stable form at 25°C.

The value of $\Delta G^{\ominus}$ is related to the equilibrium constant K for a process by the equation

$$\Delta G^{\ominus} = -RT \ln K$$

where $\ln K$ is the natural logarithm of K. Logarithms will be covered in Section 3.5.1 on page 71; don't worry about them now and just think of $\ln K$ as a single number. The other quantities in the equation are the gas constant R and the absolute temperature T. The gas constant R has the value 8.314 J K^{-1} mol^{-1}; this appears with other physical constants in Appendix B. The absolute temperature is measured in kelvin, K (not °K!). The temperature 0°C is equivalent to 273 K, but a temperature interval of 1°C is the same as a temperature interval of 1 K. For the equilibrium

$$2\,HI_{(g)} \rightleftharpoons H_{2(g)} + I_{2(g)}$$

$\ln K = -3.922$ at an absolute temperature 717 K. We then have

$$\Delta G^{\ominus} = -8.314\,\text{J K}^{-1}\,\text{mol}^{-1} \times 717\,\text{K} \times (-3.922)$$

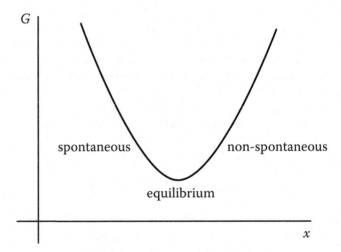

FIGURE 1.1   Graph of Gibbs free energy G against the extent of reaction x. In the spontaneous region of the graph G is falling, so ΔG is negative. ΔG then becomes zero at equilibrium, and ΔG is positive as G rises in the non-spontaneous region.

Since two negative quantities are being multiplied, the result will be positive so

$$\Delta G^{\ominus} = 8.314 \text{ J K}^{-1} \text{mol}^{-1} \times 717 \text{ K} \times 3.922$$

$$= 23.4 \times 10^3 \text{ J mol}^{-1}$$

This calculation uses the associative law of multiplication $a(bc) = (ab)c$ in that any two of the three numbers can be multiplied first, then the result multiplied by the third.

You should also notice that when combining quantities in this way we treat the units in the same way as the numbers. Since $K^{-1} = 1/K$ we have

$$\frac{1}{K} \times K = 1$$

so the unit K isn't present in the final answer.

1.2.4 DIVISION

The rules for division are similar to those for multiplication. If 9 is divided by 3, we write

$$9 \div 3 = \frac{9}{3} = 3$$

Similarly, if a is divided by b, we write this as

$$a \div b = \frac{a}{b}$$

If the first number is negative, we have

$$-9 \div 3 = \frac{-9}{3} = -\frac{9}{3} = -3$$

and if the second number is negative, we have

$$9 \div (-3) = \frac{9}{-3} = -\frac{9}{3} = -3$$

since we are combining two different signs and therefore obtain a negative result. In general terms, this is written as

$$\frac{-a}{b} = \frac{a}{-b} = -\frac{a}{b}$$

On the other hand, if both numbers are negative, the two like signs give a positive result, as in

$$(-9) \div (-3) = \frac{-9}{-3} = \frac{9}{3} = 3$$

Using the symbols a and b this can be expressed as

$$(-a) \div (-b) = \frac{-a}{-b} = \frac{a}{b}$$

The absolute temperature T at which equilibrium occurs in a system is given by the equation

$$T = \frac{\Delta H}{\Delta S}$$

where ΔH is the change in enthalpy and ΔS is the change in entropy.

For the condensation of hexane, we have $\Delta_{\text{condensation}} H = -28.9 \text{ kJ mol}^{-1}$ and $\Delta_{\text{condensation}} S = -84.5 \text{ J K}^{-1} \text{mol}^{-1}$. Thus

$$T = \frac{\Delta_{\text{condensation}} H}{\Delta_{\text{condensation}} S} = \frac{-28.9 \times 10^3 \text{ J mol}^{-1}}{-84.5 \text{ J K}^{-1} \text{mol}^{-1}} = 342 \text{ K}$$

Notice the conversion from kJ to J in $\Delta_{\text{condensation}} S$. This is necessary to give a consistent set of units with those in $\Delta_{\text{condensation}} S$. The prefix k denotes a multiplier of 10^3. A list of such SI (Système international d'unités) prefixes is given in Appendix A.1.

1.3 PRECEDENCE IN EQUATIONS

It can be difficult to know where to start when evaluating an expression with many terms and mathematical operations. This can be done by applying a set of rules known as the BODMAS rules. BODMAS stands for

- Brackets
- Orders
- Divide
- Multiply
- Add
- Subtract

These are self-explanatory except for "orders", which is another way of saying "raise to a power" or "take a root". For example

$$2^4 = 2 \times 2 \times 2 \times 2$$

You may have come across an alternative way of expressing these rules which uses the mnemonic, BIDMAS. In this version, I stands for "index", which is another name for a power or root. A further version BEDMAS uses E to stand for "exponent", but the meaning is also the same.

United States readers may be more familiar with PEMDAS as follows:

- Parentheses
- Exponents

- Multiply
- Divide
- Add
- Subtract

The important thing is that you remember the order of operations in whichever way you find most convenient.

For any expression, we perform the operations in the order specified by the BODMAS rules. Some examples will illustrate this.

Consider the expression

$$2 \times 3 + 4 \times 5$$

The rules tell us to perform multiplication before addition, so if we do this, we have

$$2 \times 3 + 4 \times 5 = 6 + 20$$

which gives the value of 26. Now, suppose that the original expression had been modified by the inclusion of brackets to become

$$2 \times (3 + 4) \times 5$$

we now have to evaluate the quantity in brackets first, and as this is equal to 7, we have

$$2 \times (3 + 4) \times 5 = 2 \times 7 \times 5$$

which has the value 70. This example shows us how brackets may be used to override the standard order in which we perform mathematical operations.

An expression which involves raising a number to a power is

$$3 \times 2^5 - 1$$

The BODMAS rules tell us to calculate the orders first, so the expression becomes

$$3 \times 32 - 1$$

since

$$2^5 = 32$$

and then

$$96 - 1$$

has the value 95.

An example of an expression in chemistry which requires the application of the BODMAS rules is the van der Waals equation

$$\left(p + \frac{an^2}{V^2}\right)(V - nb) = nRT$$

which relates the pressure p, volume V and absolute temperature T of an amount n of gas. The amount n is frequently, although incorrectly, known as the "number of moles". The quantities a and b are constants for a particular gas.

The van der Waals equation attempts to model the behaviour of a real gas by including additional correction terms involving the constants a and b to the basic gas equation for an ideal gas, $pV = nRT$.

We will consider how to evaluate the left-hand side of this expression for a sample of 1.00 mol of nitrogen gas occupying a volume of $22.414 \times 10^{-3} \, m^3$ at a pressure of $1.012 \, 2 \times 10^5 \, Pa$. The constants for nitrogen are $a = 0.140 \, 8 \, Pa \, m^6 \, mol^{-2}$ and $b = 3.913 \times 10^{-5} \, m^3 \, mol^{-1}$.

The first thing which is apparent from the BODMAS rules is that we need to begin with the brackets. In this case, this means that we evaluate each bracket in turn and then multiply the results together.

Bracket 1: We need to evaluate the term an^2/V^2 containing order, multiplication and division operations, and then add this to p.

$$\frac{an^2}{V^2} = \frac{0.140 \, 8 \, Pa \, m^6 \, mol^{-2} \times (1.00 \, mol)^2}{(22.414 \times 10^{-3} \, m^3)^2}$$

$$= \frac{0.140 \, 8 \, Pa \, m^6 \, mol^{-2} \times 1.00 \, mol^2}{502.39 \times 10^{-6} \, m^6} = 280.3 \, Pa$$

Section 1.6.3 will explain how a number containing a power can itself be raised to a power, but for the moment we can simply recognize that

$$V^2 = (22.414 \times 10^{-3} \, m^3) \times (22.414 \times 10^{-3} \, m^3)$$

Adding the terms in the first bracket now gives

$$p + \frac{an^2}{V^2} = 1.012 \, 2 \times 10^5 \, Pa + 280.3 \, Pa = 1.015 \, 0 \times 10^5 \, Pa$$

Bracket 2: We need to evaluate the term nb containing a multiplication operation, and subtract this from V.

$$nb = 1.00 \, mol \times 3.913 \times 10^{-5} \, m^3 \, mol^{-1} = 3.913 \times 10^{-5} \, m^3$$

$$V = 22.414 \times 10^{-3} \, m^3$$

$$V - nb = 22.414 \times 10^{-3} \, m^3 - 3.913 \times 10^{-5} \, m^3 = 2.237 \times 10^{-2} \, m^3$$

We now multiply the values of each bracket together, so that

$$\left(p+\frac{an^2}{V^2}\right)(V-nb)=1.015\,0\times10^5\,\text{Pa}\times2.237\times10^{-2}\,\text{m}^3=2.271\times10^3\,\text{Pa}\,\text{m}^3$$

1.4 REARRANGING EQUATIONS

An equation can be thought of as a balance of two sides, each consisting of a series of terms. In order to preserve this balance, whatever operation is applied to one side of the equation must also be applied to the other. This technique can be used to good effect to rearrange an equation. One reason for doing this might be to isolate one quantity so that its value can be calculated.

For example, the equation

$$x+y=25$$

can be arranged to make y the subject or, in other words, to begin with $y = \dots$ In order to do this, we would need to subtract x from both sides of the equation. This gives

$$x+y-x=25-x$$

which simplifies to give

$$y=25-x$$

By substituting actual numbers for x and y we can verify both the original and rearranged expressions. For example, setting $x = 10$ and $y = 15$ gives

$$10+15=25$$

in the original expression and

$$15=25-10$$

once it has been rearranged. Clearly, both these statements are correct.

This technique can also be applied to an expression involving multiplication such as

$$3y=2x$$

To rearrange this to the form $y = \dots$ we need to divide both sides by 3. This gives

$$\frac{3y}{3}=\frac{2x}{3}$$

or

$$y=\frac{2x}{3}$$

If we set $x = 3$ and $y = 2$, then

$$3 \times 2 = 2 \times 3$$

in the original expression and

$$2 = 2 \times \frac{3}{3}$$

once rearranged. Again, both statements are correct.

More complicated expressions contain both addition/subtraction and multiplication/division operations, such as

$$10x + 5y = 8$$

To obtain an expression of the form $y = ...$ we begin by subtracting $10x$ from either side to give

$$10x + 5y - 10x = 8 - 10x$$

or

$$5y = 8 - 10x$$

We then divide both sides by 5 to give

$$\frac{5y}{5} = \frac{8 - 10x}{5}$$

or

$$y = \frac{8 - 10x}{5}$$

This could also be written as

$$y = \frac{8}{5} - 2x$$

In chemistry the heat capacity C_p is defined as the quantity of heat supplied to produce a change of $1°$ (°C or K) in its temperature. It is frequently given as a series of terms of increasing power in the variable T, the absolute temperature. The simplest representation of this is

$$C_p = a + bT$$

where a and b are constants. It is possible to rearrange this to give T as a function of C_p. Subtracting a from both sides gives

$$C_p - a = a + bT - a$$

or

$$C_p - a = bT$$

Dividing both sides by b gives

$$\frac{C_p - a}{b} = \frac{bT}{b}$$

or

$$\frac{C_p - a}{b} = T$$

Conventionally, we would now reverse the order of the equation to give

$$T = \frac{C_p - a}{b}$$

although this does not change its meaning in any way.

1.5 FRACTIONS

A fractional quantity is written as two numbers separated by a horizontal line. The numerator (on top) is divided by the denominator (underneath).

For example, the fraction $\frac{4}{5}$ means "divide 4 by 5". This can also be written using the division sign as $4 \div 5$ and has the decimal value 0.8. To summarize

$$4 \div 5 = \frac{4}{5} = 0.8$$

Using algebraic symbols

$$\frac{a}{b}$$

means divide a by b, and may also be written as $a \div b$.

1.5.1 IDENTICAL FRACTIONS

If both the top and bottom of a fraction are multiplied by a constant, its value remains unchanged:

$$\frac{1}{2} = \frac{2 \times 1}{2 \times 2} = \frac{2}{4}$$

Using algebraic symbols for the general case:

$$\frac{a}{b} = \frac{c \times a}{c \times b} = \frac{ca}{cb}$$

Similarly, we can divide both top and bottom by the same number and the value of the fraction remains unchanged:

$$\frac{3}{9} = \frac{3/3}{9/3} = \frac{1}{3}$$

Again using symbols to represent a general case

$$\frac{ad}{bd} = \frac{ad/d}{bd/d} = \frac{a}{b}$$

This technique is frequently used when we perform the stoichiometric determination of an empirical formula. Such a determination of a compound containing only phosphorus and oxygen showed that a sample contained 1.409 mol of phosphorus P and 3.523 mol of oxygen O. To frame this in a more precise mathematical form we could write:

$$\text{amount of phosphorus } n_P = 1.409 \text{ mol}$$

$$\text{amount of oxygen } n_O = 3.523 \text{ mol}$$

The ratio of these quantities can then be expressed as the fraction

$$\frac{n_P}{n_O} = \frac{1.409 \,\text{mol}}{3.523 \,\text{mol}}$$

As a first step, the units of mol cancel to give

$$\frac{n_P}{n_O} = \frac{1.409}{3.523}$$

The top and bottom of this fraction can be divided by 1.409 to give

$$\frac{n_P}{n_O} = \frac{1.409/1.409}{3.523/1.409} = \frac{1}{2.5}$$

The top and bottom of this can each be multiplied by 2 to give

$$\frac{n_P}{n_O} = \frac{1}{2.5} = \frac{1 \times 2}{2.5 \times 2} = \frac{2}{5}$$

and hence the empirical formula of the compound is P_2O_5.

1.5.2 ADDITION AND SUBTRACTION

If we wish to add two fractions or to subtract one fraction from another, we need to rewrite them in terms of what is called a common denominator. This means that the number on the bottom of each of the two fractions needs to be the same. For example, we cannot calculate the sum

$$\frac{2}{3} + \frac{1}{4}$$

until the two fractions are rewritten in terms of a common denominator. The common denominator, in this case, is 12, obtained by multiplying 3 and 4. We can write

$$\frac{2}{3} = \frac{2 \times 4}{3 \times 4} = \frac{8}{12}$$

with the multiplying factor of 4 being chosen to ensure that the resulting denominator (bottom term) is 12.
   Also

$$\frac{1}{4} = \frac{1 \times 3}{4 \times 3} = \frac{3}{12}$$

The multiplying factor of 3 is chosen to ensure that the resulting denominator is 12.
   The sum is now easily evaluated as

$$\frac{2}{3} + \frac{1}{4} = \frac{8}{12} + \frac{3}{12} = \frac{11}{12}$$

The addition of two general fractions can be expressed as

$$\frac{a}{b} + \frac{c}{d} = \frac{ad}{bd} + \frac{bc}{bd} = \frac{ad + bc}{bd}$$

It is worth noting that it may be possible to simplify the result further, by dividing both top and bottom by the same number, as outlined previously in Section 1.5.1 on page 11.
   The van der Waals equation, met in Section 1.3 on page 6, can be rearranged to

$$p = \frac{nRT}{V - nb} - \frac{an^2}{V^2}$$

The common denominator of the two fractions on the right-hand side will be

$$V^2(V - nb)$$

We then have

$$\frac{nRT}{V-nb} = \frac{nRTV^2}{V^2(V-nb)}$$

and

$$\frac{an^2}{V^2} = \frac{an^2(V-nb)}{V^2(V-nb)}$$

The overall equation can then be written as

$$p = \frac{nRTV^2 - an^2(V-nb)}{V^2(V-nb)}$$

1.5.3 MULTIPLICATION

When two fractions are multiplied, we multiply the terms on the top and the terms on the bottom independently. For example,

$$\frac{2}{3} \times \frac{1}{4} = \frac{2 \times 1}{3 \times 4} = \frac{2}{12}$$

In this case, we can divide both top and bottom by 2, so that

$$\frac{2}{3} \times \frac{1}{4} = \frac{2\!\!\!/2}{12\!\!\!/2} = \frac{1}{6}$$

In general terms using symbols

$$\frac{a}{b} \times \frac{c}{d} = \frac{ac}{bd}$$

When the product in a chemical reaction becomes a reactant in a second chemical reaction, the two reactions are said to be coupled. Such a pair of reactions is

$$PO_4^{3-} + H_3O^+ \rightleftharpoons HPO_4^{2-} + H_2O$$

$$HPO_4^{2-} + H_3O^+ \rightleftharpoons H_2PO_4^- + H_2O$$

which can be combined to give the overall reaction

$$PO_4^{3-} + 2H_3O^+ \rightleftharpoons H_2PO_4^- + 2H_2O$$

The product HPO_4^{2-} in the first reaction is a reactant in the second reaction. The structures of the three anions involved in these reactions are shown in Figure 1.2.

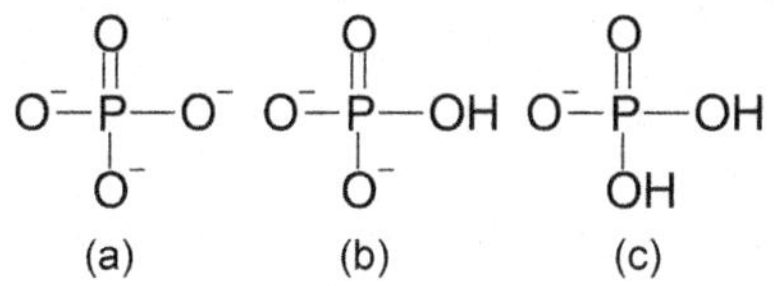

FIGURE 1.2   The anions involved in the reaction of phosphate with an acid: (a) PO_4^{3-} (b) HPO_4^{2-} (c) $H_2PO_4^-$.

The equilibrium constant K_1 for the first reaction is given by

$$K_1 = \frac{[HPO_4^{2-}][H_2O]}{[PO_4^{3-}][H_3O^+]}$$

and that for the second reaction, K_2, is given by

$$K_2 = \frac{[H_2PO_4^-][H_2O]}{[HPO_4^{2-}][H_3O^+]}$$

If K_1 and K_2 are multiplied they give the equilibrium constant K for the overall reaction

$$K = K_1K_2 = \frac{[HPO_4^{2-}][H_2O]}{[PO_4^{3-}][H_3O^+]} \times \frac{[H_2PO_4^-][H_2O]}{[HPO_4^{2-}][H_3O^+]}$$

The term $\left[HPO_4^{2-}\right]$ appears on both the top and bottom of this expression so it can be cancelled to give the overall equilibrium constant as

$$K = \frac{[H_2O][H_2PO_4^-][H_2O]^2}{[PO_4^{3-}][H_3O^+][H_3O^+]}$$

$$= \frac{[H_2PO_4^-][H_2O]^2}{[PO_4^{3-}][H_3O^+]^2}$$

since $\left[H_2O\right]\times\left[H_2O\right]=\left[H_2O\right]^2$ and $\left[H_3O^+\right]\times\left[H_3O^+\right]=\left[H_3O^+\right]^2$.

1.5.4 DIVISION

When a number is divided by a fraction, the operation we perform is to multiply by the reciprocal of the fraction. The reciprocal of a fraction is the fraction turned upside down. For example

$$\frac{2}{3} \div \frac{1}{4} = \frac{2}{3} \times \frac{4}{1} = \frac{2\times 4}{3\times 1} = \frac{8}{3}$$

In symbols, for the general case

$$\frac{a}{b} \div \frac{c}{d} = \frac{a}{b} \times \frac{d}{c} = \frac{ad}{bc}$$

The example in Section 1.5.1 on page 11 introduced the use of fractions in the determination of the empirical formula in a compound containing only phosphorus and oxygen. We can generalize this and relate it to the mass m of each element and its atomic mass M.

The amount n_p of phosphorus will be given by

$$n_p = \frac{m_p}{M_p}$$

where m_p is the mass of phosphorus in the compound and M_p is the atomic mass of phosphorus. Similarly, the amount n_O of oxygen is given by

$$n_O = \frac{m_O}{M_O}$$

where m_O is the mass of oxygen in the compound and M_O is the atomic mass of oxygen.

The ratio of the amounts of phosphorus and oxygen is then given by

$$\frac{n_p}{n_O} = \frac{m_p/M_p}{m_O/M_O} = \frac{m_p}{M_p} \times \frac{M_O}{m_O}$$

using the rule that dividing by a fraction is equivalent to multiplying by its reciprocal.

1.6 INDICES

The expression 2^4 states that the base 2 is raised to the power 4 (or has an index of 4), and this means that

$$2^4 = 2 \times 2 \times 2 \times 2$$

Here we have multiplied 2 together 4 times. In general, x^n is the base x raised to the power n, and consists of x multiplied together n times.

Some frequently encountered powers are given special names:

x^2 is called "x squared" and is equal to $x \times x$;
x^3 is called "x cubed" and is equal to $x \times x \times x$.

A little thought will show that $x^1 = x$, and we also have the special case that $x^0 = 1$. There are situations in which it is useful to express quantities with these "hidden" indices made visible.

There are a number of rules for combining numbers raised to powers. It is important to remember that such rules only apply when the quantities being combined have the same base.

1.6.1 MULTIPLICATION

When two numbers with the same base raised to powers are multiplied, we add the powers so that

$$2^2 \times 2^3 = 2^{(2+3)} = 2^5$$

In general terms, this can be written as

$$x^a \times x^b = x^{(a+b)}$$

The ionic product K_w of water is given by

$$K_w = [\text{H}^+][\text{OH}^-]$$

In the case of pure water $[\text{H}^+] = [\text{OH}^-]$, the expression becomes

$$K_w = [\text{H}^+][\text{H}^+]$$

Since $[\text{H}^+] = [\text{H}^+]^1$ this can be written as

$$K_w = [\text{H}^+]^1 \times [\text{H}^+]^1$$
$$= [\text{H}^+]^{(1+1)} = [\text{H}^+]^2$$

A similar argument can be used to show that we could alternatively write

$$K_w = [\text{OH}^-]^2$$

1.6.2 DIVISION

When one number raised to a power is divided by a number consisting of the same base raised to a power, we take the difference between the powers, so that, for example

$$\frac{2^5}{2^3} = 2^{(5-3)} = 2^2$$

In general terms,

$$\frac{x^a}{x^b} = x^{(a-b)}$$

The equilibrium constant K for the dissociation of chlorine

$$\text{Cl}_2 \rightleftharpoons 2\,\text{Cl}$$

is given by

$$K = \frac{[Cl]^2}{[Cl_2]}$$

At the point in the dissociation where

$$[Cl] = [Cl_2]$$

$$K = \frac{[Cl]^2}{[Cl]}$$

Applying the rules of indices then gives

$$K = [Cl]^{(2-1)} = [Cl]$$

At this point, it will also be true that

$$K = [Cl_2]$$

1.6.3 RAISING TO A POWER

When a number raised to a power is itself raised to a power, we multiply the powers, as in

$$(2^2)^3 = 2^{(2\times3)} = 2^6$$

This can be expressed in general terms as

$$(x^a)^b = x^{(ab)}$$

Copper (II) iodate is sparingly soluble in water according to the equation

$$Cu(IO_3)_{2(s)} \rightleftharpoons Cu^{2+}{}_{(aq)} + 2IO_3^-{}_{(aq)}$$

If the small concentration of Cu^{2+} in solution is x, it follows from the stoichiometry of the equation that the concentration of IO_3^- is $2x$. The solubility product K_{sp} is defined as

$$K_{sp} = \left[Cu^{2+}\right]\left[IO_3^-\right]^2$$
$$= x(2x)^2$$
$$= x \times 4x^2$$
$$= 4x^3$$

since $x \times x^2 = x^1 \times x^2 = x^{1+2} = x^3$

1.6.4 ROOTS

The square root of the number 16 is 4. This means that 4 multiplied by itself gives 16.

We write this mathematically as

$$\sqrt{16} = 4$$

or using an index as

$$16^{\frac{1}{2}} = 4$$

It is also possible to calculate other roots, so that the root raised to the appropriate power gives back the original number. For example, we can write that

$$\sqrt[3]{8} = 2$$

which states that the cube root of 8 is 2, and it follows that

$$2^3 = 8$$

Another way of expressing the cube root of 8 is as

$$\sqrt[3]{8} = 8^{\frac{1}{3}}$$

and, in general, when we take the root of degree b of a number x raised to power a the index of x is a divided by b, so that

$$\sqrt[b]{x^a} = x^{\frac{a}{b}}$$

Quantities raised to fractional indices occur frequently in kinetics. An example is the decomposition of ethanal (Figure 1.3) to methane and carbon dioxide:

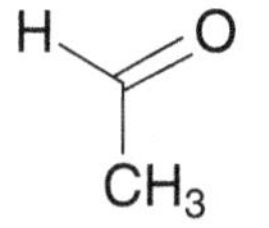

$$CH_3CHO \rightarrow CH_4 + CO$$

for which the rate of reaction is equal to $\left[CH_3CHO \right]^{\frac{3}{2}}$.

FIGURE 1.3   The chemical structure of ethanal.

1.6.5 NEGATIVE POWERS

A negative power denotes a reciprocal quantity, i.e. "one over" the number. For example

$$2^{-3} = \frac{1}{2^3}$$

In general, algebraic terms

$$x^{-n} = \frac{1}{x^n}$$

For the equilibrium dissociation of hydrogen chloride

$$2\,HCl \rightleftharpoons H_2 + Cl_2$$

the equilibrium constant K will be given by

$$K = \frac{[H_2][Cl_2]}{[HCl]^2}$$

Since

$$\frac{1}{x^2} = x^{-2}$$

this can be rewritten as

$$K = [H_2][Cl_2][HCl]^{-2}$$

1.7 STANDARD FORM

Any number can be written as the product of a number between 1 and 10 and the number 10 raised to an integral power. For example

$$2907 = 2.907 \times 10^3$$

$$0.000\,46 = 4.6 \times 10^{-4}$$

FIGURE 1.4   The chemical structure of sulfuryl chloride.

For a number less than 1 the power of 10 will be negative. To convert from a standard form note that a negative power, 10^{-n}, means to move the decimal point n places to the left, whereas a positive power, 10^n, means to move the decimal point n places to the right. In both cases, we need to insert an appropriate number of zeros to act as placeholders.

For the reaction for the decomposition of sulfuryl chloride (Figure 1.4) to sulfur dioxide and chlorine:

$$SO_2Cl_{2(g)} \rightarrow SO_{2(g)} + Cl_{2(g)}$$

the rate constant k has the value $2.20 \times 10^{-5}\,s^{-1}$. Moving the decimal point 5 places to the left gives $0.000\,022\,0\,s^{-1}$ once the necessary zeros are inserted.

The half-life $t_{1/2}$ for this reaction is $3.15 \times 10^4\,s$. The positive value of the power indicates that we move the decimal place 4 places to the right; this gives $31\,500$ seconds.

The half-life of a reaction is the time taken for the concentration of reactant to fall to half its initial value, as shown in Figure 1.5.

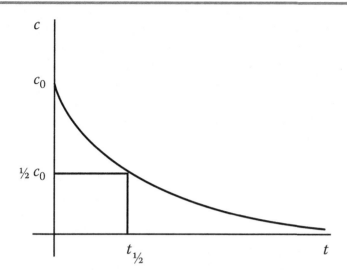

FIGURE 1.5 Graph of concentration c against time t for a reaction. The time taken for the concentration to fall to half of its initial value c_0 is the half-life $t_{\frac{1}{2}}$.

EXERCISES

1. Calculate
   a. $3 + (-1)$ b. $-7 + 2$ c. $-4 + (-2)$ d. $-3 + (-2) + (-1)$

2. Calculate
   a. $4 - 6$ b. $5 - (-2)$ c. $-6 - (-1)$ d. $-1 - (-3) - (-2)$

3. Calculate
   a. 4×7 b. $6 \times (-2)$ c. $-3 \times (-7)$ d. $-2 \times (-3) \times (-4)$

4. Calculate
   a. $\dfrac{28}{4}$ b. $\dfrac{21}{-3}$ c. $\dfrac{-25}{5}$ d. $\dfrac{-81}{-9}$

5. Evaluate
   a. $3^2 - 18/3 + 2 \times 3$ b. $(3^2 - 18)/(3 + 2 \times 3)$
   c. $3^2 - 18/(3 + 2 \times 3)$ d. $3^2 - (18/3 + 2) \times 3$

6. Rearrange the following equations to make y the subject, i.e. so that they begin with $y = ...$
   a. $x - y = 10$ b. $3x + y = 8$
   c. $4x - 2y = 10$ d. $3xy = 6$

7. Simplify each of the following to a single term in the appropriate base
   a. $3^4 \times 3^2$ b. $5^3 \times 5^{-4}$ c. $x^2 \times x^4$ d. $y^3 \times y^{-2}$

8. Simplify each of the following to a single term in the appropriate base
   a. $6^3 \div 6^4$ b. $2^4 \div 2^2$ c. $x^2 \div x^3$ d. $y^4 \div y^{-2}$

9. Simplify each of the following to a single term in the appropriate base
   a. $(4^2)^6$ b. $(3^4)^3$ c. $(x^3)^4$ d. $(y^{-2})^6$

10. Write each of the following in terms of a fractional index
    a. $\sqrt{3^5}$ b. $\sqrt[4]{6^2}$ c. $\sqrt[3]{x^5}$ d. $\sqrt[5]{y^{-3}}$

11. Write the following expressions to eliminate the fraction
    a. $\dfrac{1}{4^3}$ b. $\dfrac{1}{8^6}$ c. $\dfrac{3}{x^5}$ d. $\dfrac{2}{y^{-2}}$

12. Simplify the following fractions
    a. $\dfrac{12}{15}$ b. $\dfrac{6}{9}$ c. $\dfrac{20}{25x}$ d. $\dfrac{8x}{12y}$

13. Determine the following in their simplest fractional form
    a. $\dfrac{4}{7}+\dfrac{3}{4}$ b. $\dfrac{7}{8}+\dfrac{3}{4}$ c. $\dfrac{2}{3}-\dfrac{5}{9}$ d. $\dfrac{1}{2}-\dfrac{1}{3}$

14. Determine the following in their simplest fractional form
    a. $\dfrac{7}{8}\times\dfrac{3}{4}$ b. $\dfrac{2}{3}\times\dfrac{8}{9}$ c. $\dfrac{5}{6}\times\dfrac{2}{7}$ d. $\dfrac{5}{6}\times\dfrac{3}{8}$

15. Determine the following in their simplest fractional form
    a. $\dfrac{3}{4}\div\dfrac{5}{6}$ b. $\dfrac{2}{3}\div\dfrac{8}{9}$ c. $\dfrac{3}{5}\div\dfrac{2}{3}$ d. $\dfrac{5}{9}\div\dfrac{3}{8}$

16. Express the following numbers in standard form
    a. 36 387   b. 2 238 000 c. 0.000 176   d. 0.015 26

17. Rewrite the following numbers without using a power of 10
    a. 3.62×10^7 b. 1.66×10^4 c. 1.27×10^{-5} d. 3.948×10^{-3}

18. Calculate
    a. $-\dfrac{3}{4}+\left(-\dfrac{8}{9}\right)$ b. $\dfrac{1}{2}\times\left(-\dfrac{3}{8}\right)$

    c. $\dfrac{7}{8}-\left(-\dfrac{8}{9}\right)$ d. $\dfrac{2}{3}\div\left(-\dfrac{3}{4}\right)$

19. Evaluate
    a. $2^3{-}6/(-2) + 3 \times (-4)$ b. $4^2 + (-12)/(-3){-}2 \times (-6)$
    c. $(-5)^3{-}(-8)/(-2) - 4 \times (-3)$ d. $3^4{-}(-12)/4 - 3 \times (-6)$

20. Evaluate

a. $\left(-\dfrac{1}{2}\right)^2 + \left(-\dfrac{3}{8}\right) - \dfrac{2}{3} \times \left(-\dfrac{1}{2}\right)$

b. $-\left(\dfrac{3}{4}\right)^2 - \left(-\dfrac{5}{8}\right) \div \left(-\dfrac{3}{4}\right)$

c. $\left(-\dfrac{2}{3}\right)^2 - \left(-\dfrac{1}{4}\right)^2 \div \left(-\dfrac{1}{2}\right)$

d. $\left(\dfrac{3}{4}\right) \times \left(-\dfrac{1}{2}\right) - \left(-\dfrac{7}{8}\right) \div \left(-\dfrac{5}{6}\right)$

PROBLEMS

1. The Born-Haber cycle for sodium chloride allows its lattice enthalpy $\Delta_L H$ to be calculated in terms of various other enthalpy changes. This leads to the equation

$$-\left(-411 \text{ kJ mol}^{-1}\right) + 107 \text{ kJ mol}^{-1} + 496 \text{ kJ mol}^{-1} + 122 \text{ kJ mol}^{-1}$$

$$+\left(-349 \text{ kJ mol}^{-1}\right) + \Delta_L H = 0$$

Calculate $\Delta_L H$ from this equation.

2. Coulomb's law gives the potential energy V in terms of the distance r between two point charges q_1 and q_2 as

$$V = \frac{q_1 q_2}{4\pi\varepsilon_o r}$$

where ε_o is a constant known as the permittivity of free space. Determine an expression for the potential energy V between a nucleus with atomic number Z having charge Ze and an electron having charge $-e$.

3. For a chemical reaction the change in Gibbs free energy ΔG is related to the change in enthalpy ΔH and the change in entropy ΔS by

$$\Delta G = \Delta H - T\Delta S$$

where T is the absolute temperature. Rearrange this equation to give an expression for T.

4. The chemical equation for catalytic methanation is

$$CO_{(g)} + 3\,H_{2(g)} \rightleftharpoons CH_{4(g)} + H_2O_{(g)}$$

with a corresponding equilibrium constant K given by

$$K = \frac{[CH_4][H_2O]}{[CO][H_2]^3}$$

Determine a simplified form of this expression when

a. $[H_2] = [H_2O]$ b. $[H_2] = 2\,[CO]$

c. $[H_2] = \frac{1}{2}\,[CO]$ d. $[CO] = [H_2]$

by eliminating $[H_2]$ from the equation.

5. The rate equation for the reaction

$$2\,NO_{2(g)} \rightarrow 2\,NO_{(g)} + O_{2(g)}$$

is

$$\frac{1}{[NO_2]} = kt + \frac{1}{[NO_2]_o}$$

where $[NO_2]_o$ is the initial concentration of NO_2 and $[NO_2]$ its concentration after time t. The rate constant k is a measure of how fast the reaction proceeds.

Rearrange this equation to express:

a. kt as a difference of two fractions;

b. kt in terms of a single fraction with denominator $[NO_2][NO_2]_o$.

6. The quantum yield ϕ for a photochemical reaction can be expressed in terms of the rate constants k for various processes as

$$\frac{1}{\phi} = \frac{k_F + k_{IC} + k_{ISC}}{k_{ISC}}\left(1 + \frac{k_P + k'_{ISC}}{k_R[R]}\right)$$

where $[R]$ is the concentration of reactant. The processes involved are fluorescence (F), internal conversion (IC) and intersystem crossing (ISC), and k_R is the bimolecular rate constant. Write an expression for $\frac{1}{\phi}$ as a single fraction.

7. The collision density Z_{AB} between two chemical species A and B is defined as

$$Z_{AB} = \sigma\left(\frac{8kT}{\pi\mu}\right)^{\frac{1}{2}} N_A^2[A][B]$$

where σ is the collision cross section, μ a quantity called the reduced mass, T the absolute temperature and $[A]$ and $[B]$ the concentrations of A and B, respectively. The other quantity in this equation is the Boltzmann constant, k. Rearrange this equation to obtain an expression for μ.

8. Write the following physical constants and conversion factors in standard form:

a. $c = 299\,800\,000$ m s^{-1}

b. $R = 10\,970\,000$ m^{-1}

c. $F = 96\ 490\ C\ mol^{-1}$
d. $p = 101\ 325\ Pa$
e. $V_0 = 0.022\ 414\ 1\ m^3$

9. Rewrite the following quantities without the use of standard form, i.e. as a decimal quantity:
   a. $V = 2.196 \times 10^3\ m^3$
   b. $p = 1.86 \times 10^4\ Pa$
   c. $E = 3.142 \times 10^{-3}\ V$
   d. $k = 3.84 \times 10^5\ dm^3\ mol^{-1}\ s^{-1}$
   e. $n = 8.544 \times 10^{-5}\ mol$

10. The wavenumbers $\bar{\upsilon}$ of lines in the hydrogen-like atom spectrum are given by

$$\bar{\upsilon} = RZ^2 \left(\frac{1}{n_1^2} - \frac{1}{n_2^2} \right)$$

where n_1 is the lower level of the transition, n_2 the upper level, R the Rydberg constant and Z the nuclear charge. Obtain expressions for $\bar{\upsilon}$ when
   a. $n_1 = 1$ and $n_2 = 2$
   b. $n_1 = 2$ and $n_2 = 3$
   c. $n_2 = 2n_1$

11. The heat capacity C_p of a substance at constant pressure can be expressed in terms of the absolute temperature T as

$$C_p = a + bT + \frac{c}{T^2}$$

where a, b and c are constants. For graphite, the constants take the values $a = 16.86\ J\ K^{-1}\ mol^{-1}$, $b = 4.77 \times 10^{-3}\ J\ K^{-2}\ mol^{-1}$ and $c = -8.54 \times 10^5\ J\ K\ mol^{-1}$. Determine the value of C_p for graphite at 325 K.

12. The decomposition of ozone catalysed by N_2O_5 is represented by the equation

$$2\ O_3 \rightleftharpoons 3\ O_2$$

and the rate v of this reaction is given by

$$v = k[O_3]^{\frac{2}{3}}[N_2O_5]^{\frac{2}{3}}$$

where k is the rate constant.
   a. Obtain an expression for the rate when $[O_3] = [N_2O_5]$.
   b. Obtain an expression for the rate when $[O_3] = 2\ [N_2O_5]$.
   c. Obtain an expression for the rate when $[O_3] = 0.12\ mol\ dm^{-3}$ and $[N_2O_5] = 0.16\ mol\ dm^{-3}$.

Experimental Techniques

2.1 INTRODUCTION

Despite the increasing use of computers in many aspects of chemistry and the enormous advances in theoretical areas such as quantum mechanics, many chemists still perform laboratory experiments. Particularly in physical chemistry, the objective of an experiment is often to collect numerical data which can then be analysed to determine the value of some physical quantity. It is important that you are able to manipulate this sort of data as required. At first sight, this may seem trivial, but in some ways, the appropriate treatment of data is one of the most difficult areas of numerical chemistry to master. Every case seems to be different, and it is only by gaining experience in several calculations that you will become confident in performing this kind of data analysis.

Note that you will often come across the term "errors" when experimental quantities are being discussed. I prefer the term "uncertainties" as this indicates an intrinsic property of data, rather than suggesting that a mistake has been made in its measurement or analysis.

We will begin our consideration of experimental uncertainties by considering an example of how these are expressed. First of all, we need to consider how experimental uncertainties are expressed.

A solution having a concentration of $0.002\,10\,\mathrm{mol\,dm^{-3}}$ of bromine in tetrachloromethane was found to have an absorbance of 0.796 ± 0.003. The presence of the plus and minus sign $\pm$ in this expression immediately alerts us to the fact that we are being present with a measured quantity together with its uncertainty. Performing the subtraction operation first suggests that

$$\text{minimum absorbance} = 0.796 - 0.003 = 0.793$$

while performing the addition operation suggests that

$$\text{maximum absorbance} = 0.796 + 0.003 = 0.799$$

Taking these values together, it is reasonable to state that we would expect the absorbance of this sample to fall between 0.793 and 0.799. As we will see, there are different ways of determining uncertainties, so the precise meaning of a statement like this will be slightly vague at the moment.

DOI: 10.1201/9781003043218-2

Accuracy and precision are terms which are often confused. Accuracy denotes how closely a measured value is to the true value of a quantity, while precision is an indication of how reliably a measurement can be reproduced. It is thus possible to have precise measurements which are not accurate, and accurate measurements which are not precise.

2.2 MEASUREMENT IN CHEMISTRY

Many of the measurements made in physical chemistry are of quantities which are easy to visualize. For example, a burette is used to measure volume, and a thermometer is used to measure temperature. Other measurements which you may make in a university chemistry course will be of less familiar quantities. Absorbance is measured using a spectrophotometer and electromotive force is measured using a digital voltmeter, for example. Frequently, the values read from these instruments will be used in subsequent calculations, which will usually be performed with the assistance of an electronic calculator. Such calculators do allow you to perform complex calculations easily, but unfortunately, they do not always give results which can be quoted directly to give an appropriate answer without considering the validity of the precision with which it is quoted.

A common example will serve to illustrate this, using a procedure you will have already encountered in the laboratory. Suppose you wish to standardize a solution of sodium hydroxide which has a concentration of approximately $0.1 \, mol \, dm^{-3}$. To do this, titrate $25 \, cm^3$ of the sodium hydroxide with hydrochloric acid which, you are told, has a concentration of $0.099 \, 4 \, mol \, dm^{-3}$. Suppose that $25 \, cm^3$ of sodium hydroxide is neutralized by $24.85 \, cm^3$ of hydrochloric acid. The concentration of sodium hydroxide, [NaOH], is then given by the formula

$$[NaOH] = \frac{24.85 \, cm^3 \times 0.099 \, 4 \, mol \, dm^{-3}}{25 \, cm^3}$$

Using the calculator, this gives a value for [NaOH] of $0.098 \, 803 \, 6 \, mol \, dm^{-3}$. The question we need to ask ourselves is "are we justified in quoting the final answer to this many figures?". You may guess that, in this case, the answer is "no" since there is a disparity between the number of figures in this answer and that in the original data. This is a vague statement, however, and how do we decide what number of figures would be reasonable in the final answer? To do this we need to understand the concepts of decimal places and significant figures.

2.2.1 DECIMAL PLACES

In the example above, we could quote the final value for [NaOH] to any number up to 7 decimal places. Some examples would be

$[NaOH] = 0.098 \, 8 \, mol \, dm^{-3}$ to 4 decimal places
$[NaOH] = 0.098 \, 80 \, mol \, dm^{-3}$ to 5 decimal places
$[NaOH] = 0.098 \, 803 \, 6 \, mol \, dm^{-3}$ to 7 decimal places

The number of decimal places is the number of digits which appear after the decimal point.

2.2.2 SIGNIFICANT FIGURES

The number of significant figures in each expression is, however, quite different from the number of decimal places. To count these, we need to apply the following rules:

1. Ignore any leading zeros (such as those in 0.001, for example).
2. Include any zeros "inside" the number (such as the second zero in 0.101, for example).
3. Use your judgement to decide whether trailing zeros are significant or not. This is more likely to be the case when they appear after the decimal point. If a number has been expressed as 37.00, for example, it is unlikely that the writer intended to write the more approximate 37, at least if they had thought about what they were doing! Similarly, writing 0.003 700 with four significant figures implies greater precision than writing 0.003 7 with only two.
4. All non-zero digits are significant.

So, for the expressions of sodium hydroxide concentration above, we now have

$[NaOH] = 0.098\ 8\ mol\ dm^{-3}$ to 3 significant figures, since the leading zeros are not significant.

$[NaOH] = 0.098\ 80\ mol\ dm^{-3}$ to 4 significant figures, since the leading zeros are not significant and the trailing zero is significant as it increases the precision to which the number is specified

$[NaOH] = 0.098\ 803\ 6\ mol\ dm^{-3}$ to 6 significant figures, since the leading zeros are not significant and the enclosed zero is significant.

Now, consider what happens if we wish to express the concentration to five significant figures. We might expect to obtain a value of $0.098\ 803\ mol\ dm^{-3}$, but if we look at the first digit immediately to the right of this that we have deleted, we find that it is a 6. This means that the value we require is, in fact, closer to $0.098\ 804\ mol\ dm^{-3}$ than it is to $0.098\ 803\ mol\ dm^{-3}$, and we need to "round up" the last digit to give the higher value.

When quoting a value to a specified number of decimal places or significant figures, always look at the first digit you are planning to delete. If it is between 5 and 9 round up the previous digit; otherwise leave the previous digit as it is.

One case in which the application of the rules of significant figures can seem a little strange is when you are asked to round a number greater than 10 to fewer figures than there are digits present before the decimal point, such as 1265 to 2 significant figures. In such a case, we need to fill out the remaining places with zeros; in this case, we need to round successively. 1265 rounds to 1270 which in turn rounds to 1300. Remember that subsequently the zeros in this number are not significant.

Worked Example 2.1

Write each of the following quantities to (a) three significant figures and (b) two decimal places:

    a. a volume of 17.927 cm^3
    b. an energy change of 396.748 kJ mol^{-1}
    c. a rate constant of 1.084 6 dm^3 mol^{-1}s^{-1}
    d. the gas constant 0.082 057 8 dm^3 atm K^{-1} mol^{-1}

CHEMICAL BACKGROUND

    a. We have already seen that the measurement of volume is important in chemistry in volumetric analysis. Volume changes are also important in thermodynamics where much emphasis is placed on the behaviour of gases.
    b. Energy changes accompany chemical reactions and can be predicted using thermodynamic methods. They can be measured experimentally using one of the various types of calorimeter available, such as that shown in Figure 2.1.
    c. The rate constant determines how fast a particular chemical reaction will take place. You may have noticed that the units can be thought of as "per concentration per time", and it is the value of the concentration which often governs the rate of a chemical reaction at a given temperature. As we will see in Chapter 5, the rate constant has different units depending on the type of reaction to which it refers.
    d. We usually express the gas constant in units of J K^{-1}mol^{-1}, but the form given in this example can be useful when pressure values are given in units of atm. The gas constant appears in many branches of physical chemistry, and not just those concerned with gases!

Solution to Worked Example 2.1

    a. Counting the first 3 significant figures is straightforward and gives a value of 17.9 cm^3. There is no need to round up because the next digit is 2. To 2 decimal places, we obtain 17.92 cm^3, but a

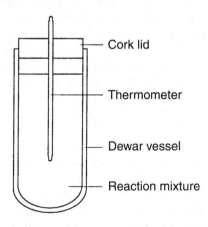

FIGURE 2.1   Schematic diagram of a Dewar calorimeter.

7 has been deleted (which falls between 5 and 9), so we round up the last digit to give 17.93 cm³.

b. The first 3 significant figures take us as far as the decimal point, but the first digit deleted is 7 (falling between 5 and 9) so we round 6 up to 7 to give 397 kJ mol⁻¹. To 2 decimal places we have 396.74 kJ mol⁻¹, but the first digit deleted is 8 so we round up to get 396.75 kJ mol⁻¹.

c. The first 3 significant figures give 1.08 dm³ mol⁻¹ s⁻¹ since the zero inside the number is included as a significant digit. The first digit deleted is 4 (which falls between 0 and 4) so no rounding is required. The result is the same as 1.08 dm³ mol⁻¹ s⁻¹ when we round to 2 decimal places.

d. We ignore the leading zeros and count 3 significant figures to give 0.082 0 dm³ atm K⁻¹ mol⁻¹. However, the first digit deleted is 5, so we round up to give 0.082 1 dm³ atm K⁻¹ mol⁻¹. Counting 2 decimal places gives 0.08 dm³ atm K⁻¹ mol⁻¹, and as a 2 has been deleted (between 0 and 4) no additional rounding is required.

2.2.3 COMBINING QUANTITIES

Having decided how many significant figures are appropriate for each quantity involved in a calculation, we are in a position to determine the appropriate number of digits to which to quote our final calculated value. To do this we need to apply the following rules:

1. When adding or subtracting numbers, the answer should be given to the smallest number of decimal places. For example, we would write $1.23 + 4.5 = 5.7$ rather than $1.23 + 4.5 = 5.73$.

2. When multiplying numbers, or dividing one number by another, the answer should be given to the smallest number of significant figures. For example, we would write

$$\frac{7.432}{2.1} = 3.5$$

rather than

$$\frac{7.432}{2.1} = 3.539$$

since a quantity given to 4 significant figures is being divided by one given to 2 significant figures. The result is quoted to the lesser number of significant figures which here is 2.

Worked Example 2.2

Find the value of the following expressions, giving the appropriate number of figures in your answer.

a. The difference between burette readings of 36.35 and 11.2 cm³.
b. The density of a solution calculated from the formula

$$\text{density} = \frac{\text{mass}}{\text{volume}} = \frac{17.098\,\text{g}}{6.2\,\text{cm}^3}$$

c. The concentration of a solution given by the expression

$$\frac{25.15\,\text{cm}^3 \times 0.104\,\text{mol dm}^{-3}}{25.00\,\text{cm}^3}$$

CHEMICAL BACKGROUND

a. This is an example of bad laboratory practice. The burette should always be read to the same degree of precision, which would typically be to ±0.05 cm³. This is true even if the starting value was exactly zero, so this would be given as 0.00 cm³ rather than 0.0 cm³.
b. The density of a substance is found by dividing the mass of a sample by its volume. One example of doing this in the laboratory, for solutions, is an experiment to determine partial molar volumes. These are discussed in Section 4.8 on pages 137 to 142.
c. This is another example of the standardization procedure met earlier in this section. It is important to realize that the division by 25.00 cm³ refers to a pipette volume, so the trailing zeros are important to take into account the precision of the pipette.

Solution to Worked Example 2.2

a. The subtraction (36.35 − 11.2) cm³ on a calculator gives the result 25.15 cm³. However, we can see that the second of the two values is given to only 1 decimal place, so the final answer would be 25.1 cm³. However, digit 5 has been deleted, so we need to round up the final value to give 25.2 cm³.
b. Using the calculator gives an answer of 2.757 741 9 g cm⁻³ which contains more figures than we require. If we look at the two quantities given in the question, we see that they contain 5 and 2 significant figures, respectively. Giving the answer to 2 significant figures gives 2.7 g cm⁻³, but we need to remember that we have deleted the digit 5 so we need to round up the answer to 2.8 g cm⁻³.
c. The calculator again gives a large number of figures, and we obtain an initial answer of 0.104 624 mol dm⁻³. Looking at the three quantities in the question shows that they are given to 4, 3 and 4 significant figures, respectively. We include in this count the zero inside the second quantity and the trailing zeros in the pipette volume. The answer therefore needs to be given to 3 significant figures which give 0.104 mol dm⁻³. The first digit deleted is 6, so once again we need to round up so the final answer is given as 0.105 mol dm⁻³.

2.3 STOICHIOMETRIC CALCULATIONS

In the previous section, we used calculations based on volumetric analysis as examples. The reaction between sodium hydroxide and hydrochloric acid takes place in a simple 1:1 ratio, so we did not have to consider this aspect of the reaction any further. Many chemical reactions are less straightforward than this, but it is useful to be able to calculate the amount or mass of products from a given amount or mass of reactants.

2.3.1 MULTIPLICATION AND DIVISION BY AN INTEGER

In previous examples of calculations, we have been dealing with real numbers where it has been necessary to determine the number of significant figures (or decimal places) in each value included. However, when multiplying or dividing by an integer, we do not need to consider how many significant figures it has. This is because integers are exact numbers so that 4 could be thought of as 4.000 000 0....., for example, to ensure that it is treated as having more significant figures than any other quantity in the calculation.

Some simple examples of this are

$$3 \times 5.983 = 17.95 \text{ given to the 4 significant figures of 5.983}$$

and

$$\frac{3.16}{5} = 0.632 \text{ given to the 3 significant figures of 3.16.}$$

However, we need to be sure that these are integers and not rounded numbers. This is the case for stoichiometric coefficients.

Worked Example 2.3

In the reaction

$$Fe_2O_{3(s)} + 3\ CO_{(g)} \rightarrow 2\ Fe_{(l)} + 3\ CO_{2(g)}$$

calculate the mass of iron produced when 503 g of carbon monoxide is passed over an excess of Fe_2O_3. The molecular mass of CO is 28.01 g mol^{-1} and the atomic mass of iron is 55.85 g mol^{-1}.

CHEMICAL BACKGROUND

This equation represents the reaction which takes place in a blast furnace (Figure 2.2), where iron is extracted from iron ore by reduction with carbon monoxide. Coke is burned to form carbon dioxide which then reacts with more hot coke to form carbon monoxide.

Unlike the previous examples, we have met, where equal number of molecules of each reactant combine, here we see that one molecule of iron oxide combines with three molecules of

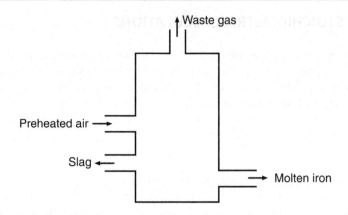

FIGURE 2.2   Schematic diagram of a blast furnace.

carbon monoxide. The fact that these numbers are still integers is a consequence of Dalton's atomic theory expressed in the law of multiple proportions which was formulated in 1803.

Solution to Worked Example 2.3

It is important to realize that the coefficients which appear in this equation refer to the relative numbers which combine in the reaction. These are known as the amounts of each species and are measured in moles. Using the relationship

$$\text{amount} = \frac{\text{mass}}{\text{molecular mass}}$$

and substituting the given values into this equation, we have

$$\text{amount of CO} = \frac{503 \text{ g}}{28.01 \text{ g mol}^{-1}}$$

and

$$\text{amount of Fe} = \frac{\text{mass of Fe}}{55.85 \text{ g mol}^{-1}}$$

From the reaction equation, we see that the ratio of the amount of CO:amount of Fe is 3:2, which is equivalent to the equation

$$\frac{\text{amount of CO}}{\text{amount of Fe}} = \frac{3}{2}$$

We can rearrange this, using the rules given in Section 1.4 on pages 9 to 11, to give

$$\text{amount of Fe} = \left(\frac{2}{3}\right) \times \text{amount of CO}$$

Substituting in our expression for the amounts of each of these gives

$$\frac{\text{mass of Fe}}{55.85 \text{ g mol}^{-1}} = \left(\frac{2}{3}\right) \times \frac{503 \text{ g}}{28.01 \text{ g mol}^{-1}}$$

This equation can be rearranged by multiplying both sides by 55.85 g mol^{-1} to give the final expression

$$\text{mass of Fe} = 55.85 \text{ g mol}^{-1} \times \left(\frac{2}{3}\right) \times \frac{503 \text{ g}}{28.011 \text{ g mol}^{-1}}$$

Evaluating this on a calculator gives the value 668.607 57 g. We can ignore the exact integers 2 and 3 in considering the appropriate number of significant figures and are then left with 4, 3 and 4 figures respectively in the remaining quantities. It is therefore appropriate to give the answer to 3 significant figures, so our final answer should be 669 g, as we need to round up the final digit as the first digit deleted is 6.

It is worth noting that the calculator was only used at the very end of this process. Doing this reduces the chances of mistakes occurring in the intermediate steps affecting the final result.

Also, note that if we had converted $\frac{2}{3}$ to a decimal value earlier in the calculation we might have been unable to determine the correct number of significant figures in the final answer.

Worked Example 2.4

What volume of sulfuric acid of concentration 0.050 2 mol dm^{-3} is required to neutralize 25.00 cm^3 of 0.099 5 mol dm^{-3} sodium hydroxide?

CHEMICAL BACKGROUND

Sulfuric acid is one of the chemicals produced on a large industrial scale. It is used in making fertilizers and as a catalyst in the refining of petroleum. It is known as a dibasic acid because it contains two hydrogen atoms which may dissociate from the molecule to give the hydrogen ions which confer acidity (Figure 2.3). In hydrochloric acid, there is only one such hydrogen and the acid is known as a monobasic acid.

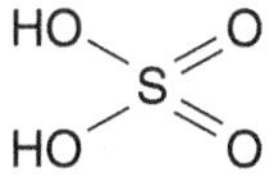

FIGURE 2.3 Structure of sulfuric acid.

Solution to Worked Example 2.4

The equation for the neutralization reaction taking place is

$$2NaOH + H_2SO_4 \rightarrow Na_2SO_4 + 2H_2O$$

so we see that only 1 mol of sulfuric acid is required for every 2 mol of sodium hydroxide. Since the ratio amount of sodium hydroxide: amount of sulfuric acid is 2: 1 we can write

$$\frac{\text{amount of sodium hydroxide}}{\text{amount of sulfuric acid}} = \frac{2}{1}$$

and then substitute the values into this equation to give

$$\frac{25.00 \, \text{cm}^3 \times 0.099\,5 \, \text{mol dm}^{-3}}{\text{volume of sulfuric acid} \times 0.050\,2 \, \text{mol dm}^{-3}} = \frac{2}{1}$$

This rearranges to give the equation

$$\text{volume of sulfuric acid} = \frac{25.00 \, \text{cm}^3 \times 0.099\,5 \, \text{mol dm}^{-3}}{0.050\,2 \, \text{mol dm}^{-3} \times 2}$$

which, on using a calculator, yields a volume of 24.775 896 42 cm³. To determine the number of significant figures in this answer, we have to realize that the volume of sodium hydroxide will have been determined by a pipette and so the trailing zeros given in the quantity 25.00 cm³ will be significant. Ignoring the integer value of 2, we are left with combining 25.00 cm³ which has at least 4 significant figures, 0.099 5 mol dm⁻³ which has 3 significant figures (having ignored leading zeros) and 0.050 2 mol dm⁻³ which has 3 significant figures (ignoring the leading zero but including the one 'inside' the number), so our answer should be given to 3 significant figures. Rounding up the third of these leads to a final value for the volume of 24.8 cm³.

2.4 UNCERTAINTY IN MEASUREMENT

So far, we have addressed the question of how many figures are justified when quoting the numerical value of a particular quantity. In other words, we have seen how to specify the appropriate precision of a physical quantity. We now need to consider how we can judge the accuracy of our value.

The object of many physical chemistry experiments is to obtain the value of some numerical quantity. Examples are the rate constant for a reaction, the dissociation constant of an acid or the enthalpy change accompanying a reaction. Frequently, in student work, the accepted values for these quantities are well documented and a student can assess the uncertainty in the determination of a particular value by consulting the literature.

In genuine research work, this is not the case. You might be studying a new reaction or using a new technique, and have no idea what the exact value is going to be. In this case, it is vital to be able to estimate the uncertainty associated with such a value. Even if the value is known, you need to be able to state whether your experimental value agrees with the literature value within the bounds of experimental uncertainty.

There are a number of approaches to the determination of experimental uncertainties. When carrying out research, the determination of a physical quantity would be repeated many times, and statistical methods can then be used as described later in this section. You are more likely to be in the position of having a fixed number of hours to perform an experiment, after which the demands of the course dictate that you move on to new work. The results of the class may be pooled to allow a statistical treatment but, alternatively, you will be left with a single determination of your value on which you need to estimate the uncertainty. In this case, you will have to estimate the probable uncertainty; we will see shortly that there are two ways in which this may be done.

2.4.1 TYPES OF UNCERTAINTY

You may decide that you can read a burette to $\pm 0.05 \, cm^3$, a balance to $\pm 0.000 \, 1g$ or a digital voltmeter to $\pm 0.001 \, V$ for example. As we saw in Section 2.1, we use the $\pm$ symbol in the specification of uncertainties to show that they have an equal chance of being too high or too low by this amount.

You need to be able to distinguish between three definitions of uncertainties. To illustrate these, suppose you have estimated that the uncertainty on a quantity X is ΔX. As in Section 1.2.1, the Greek letter Δ is used to represent a difference in values. We can then define

- the absolute uncertainty as ΔX

- the fractional uncertainty as $\dfrac{\Delta X}{X}$

- the percentage uncertainty as $100 \times \dfrac{\Delta X}{X}$

Worked Example 2.5

An actual volume of $25.00 \, cm^3$ is measured as $25.15 \, cm^3$. What is (a) the absolute uncertainty, (b) the fractional uncertainty and (c) the percentage uncertainty in this quantity?

CHEMICAL BACKGROUND

The uncertainty involved in this example is rather more than you should hope to achieve in practice! It helps to ensure that the burette is mounted vertically and that the effects of parallax are avoided when readings are taken, as shown in Figure 2.4.

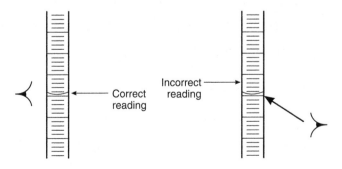

FIGURE 2.4   Effect of parallax when reading a burette.

Solution to Worked Example 2.5

a. The absolute uncertainty is ±0.15 cm³; we use the ± sign because we would not generally know in which direction the uncertainty was.

b. The fractional uncertainty is

$$\pm\frac{0.15\,\text{cm}^3}{25.00\,\text{cm}^3} = \pm\,0.006$$

c. The percentage uncertainty will be 100 times the fractional uncertainty, which is ±0.6%.

2.4.2 COMBINING UNCERTAINTIES

The determination of the value of a physical quantity in an experiment will usually involve several measurements followed by one or more calculations involving these quantities. As these are combined using mathematical operations such as addition, subtraction, multiplication and division, the corresponding uncertainties are also combined using a set of rules. Suppose that we have measured two quantities X and Y and have estimated the uncertainties on them to be ΔX and ΔY respectively.

2.4.2.1 DETERMINING THE MAXIMUM POSSIBLE UNCERTAINTY

This can be thought of as the "common sense" approach. We don't need to remember any formulae, but we do need to be vigilant at every stage of a calculation.

a. If X and Y are added, we have
Maximum value of $(X+Y)=(X+\Delta X)+(Y+\Delta Y)$
Minimum value of $(X+Y)=(X-\Delta X)+(Y-\Delta Y)$

b. If Y is subtracted from X, we have
Maximum value of $(X-Y)=(X+\Delta X)-(Y-\Delta Y)$
Minimum value of $(X-Y)=(X-\Delta X)-(Y-\Delta Y)$

c. If X and Y are multiplied, we have
Maximum value of $XY=(X+\Delta X)\times(Y+\Delta Y)$
Minimum value of $XY=(X-\Delta X)\times(Y-\Delta Y)$

d. If X is divided by Y, we have

$$\text{Maximum value of }\frac{X}{Y}=\frac{X+\Delta X}{Y-\Delta Y}$$

$$\text{Minimum value of }\frac{X}{Y}=\frac{X-\Delta X}{Y+\Delta Y}$$

Note that the maximum value of X/Y is obtained by dividing the maximum value of X by the minimum value of Y, while the minimum value of X/Y is obtained by dividing the minimum value of X by the maximum value of Y.

In each case, the uncertainty is given as

$0.5 \times$ (maximum value $-$ minimum value).

2.4.2.2 DETERMINING THE MAXIMUM PROBABLE UNCERTAINTY

a. If X and Y are added, or one is subtracted from the other, the absolute uncertainty of the result will be

$$\sqrt{(\Delta X)^2 + (\Delta Y)^2}$$

b. If X and Y are multiplied, or one is divided by the other, the fractional uncertainty of the result will be

$$\sqrt{\left(\frac{\Delta X}{X}\right)^2 + \left(\frac{\Delta Y}{Y}\right)^2}$$

Following on from this, if X is raised to the power n, the fractional uncertainty on the product will be

$$\frac{\Delta X}{X}\sqrt{n}$$

The percentage uncertainty can then be obtained by multiplying the fractional uncertainty by 100.

Worked Example 2.6

The initial and final burette readings in a titration are (5.00 ± 0.05) cm^3 and (31.85 ± 0.05) cm^3 respectively. What are the maximum possible and maximum probable uncertainties in the titre?

CHEMICAL BACKGROUND

While it is probably more usual to refill a burette to the zero mark before starting each titration, there is an advantage to not doing this. Using a greater length of the burette will help to average out variations in the bore. It is important, however, to ensure that you do not run out of solution before the end point has been reached!

Burettes are manufactured and calibrated in accordance with the International Organization for Standardization Standards. Class A burettes are made from borosilicate and retain their accuracy for longer than class B burettes made from soda lime.

Solution to Worked Example 2.6

The value of the titre is found by subtracting the initial volume from the final volume, so we have

$$\text{Titre} = (31.85 - 5.00) \text{ cm}^3 = 26.85 \text{ cm}^3$$

We begin by calculating the maximum possible uncertainty. We have

$$\text{Maximum titre} = (31.90 - 4.95) \text{ cm}^3 = 26.95 \text{ cm}^3$$
$$\text{Minimum titre} = (31.80 - 5.05) \text{ cm}^3 = 26.75 \text{ cm}^3$$

with an uncertainty of

$$0.5 \times (26.95 - 26.75) \text{ cm}^3 = 0.5 \times 0.20 \text{ cm}^3 = 0.10 \text{ cm}^3$$

We calculate the maximum probable uncertainty as

$$\sqrt{(\Delta X)^2 + (\Delta Y)^2} = \sqrt{(0.05 \text{cm}^3)^2 + (0.05 \text{cm}^3)^2} = 0.07 \text{cm}^3$$

Note that the maximum probable uncertainty is less than the maximum possible uncertainty. This will generally be the case; the formula for the maximum probable uncertainty takes account of the fact that an uncertainty will not normally have its maximum value. From this point, we will calculate only the maximum probable uncertainty.

It is also worth noting that the uncertainty value of 0.07 cm³ has been obtained independently of the titre value, and so is valid for any titration in which the burette has been consistently read to ±005 cm³. We would express the final titre as (26.85±0.07) cm³. The brackets are used here to show that the units apply to both the titre value and its uncertainty but, by convention, they may be omitted in which case the titre would be given as 26.85±0.07 cm³.

Worked Example 2.7

Calculate the solubility of magnesium ethanedioate from the following information. A volume of 14.85±0.07 cm³ of acidified potassium permanganate solution of concentration 0.005 0 ± 0.000 05 mol dm⁻³ is required to oxidize 20.0±0.1 cm³ of a saturated solution of the oxalate. This reaction can be represented by the equation

$$5 \text{ C}_2\text{O}_4^{2-} + 2 \text{ MnO}_4^- + 16 \text{ H}^+ \rightarrow 10 \text{ CO}_2 + 8 \text{ H}_2\text{O} + 2 \text{ Mn}^{2+}$$

CHEMICAL BACKGROUND

This is an example of a redox titration in which electron transfer is involved. The permanganate or manganate (VII) ion MnO_4^- is reduced to manganese (II) ions according to the equation

$$\text{MnO}_4^- + 8 \text{ H}^+ + 5 \text{ e}^- \rightarrow \text{Mn}^{2+} + 4 \text{ H}_2\text{O}$$

while the ethanedioate ion $\text{C}_2\text{O}_4^{2-}$ is oxidized to carbon dioxide by the reaction

$$\text{C}_2\text{O}_4^{2-} \rightarrow 2\text{CO}_2 + 2 \text{ e}^-$$

Solution to Worked Example 2.7

The only information which needs to be extracted from this equation is that the ratio amount of ethanedioate: amount of permanganate is $5 : 2$. This gives us the equation

$$\frac{\text{amount of ethanedioate}}{\text{amount of permanganate}} = \frac{5}{2}$$

Substituting values for the concentration and volume leads to

$$\frac{20.0\,\text{cm}^3 \times [\text{C}_2\text{O}_4^{2-}]}{14.85\,\text{cm}^3 \times 0.005\,0\,\text{mol dm}^{-3}} = \frac{5}{2}$$

which can be rearranged to give

$$[\text{C}_2\text{O}_4^{2-}] = \left(\frac{5}{2}\right) \times \frac{14.85\,\text{cm}^3 \times 0.005\,0\,\text{mol dm}^{-3}}{20.0\,\text{cm}^3}$$

The use of a calculator gives the value $9.3 \times 10^{-3}\,\text{mol dm}^{-3}$, applying the rules we have learned for significant figures in Section 2.2.2 and realising that the trailing zero in the value of the permanganate concentration is significant to give an answer to 2 significant figures.

If we now look at the equation we used to evaluate $\left[\text{C}_2\text{O}_4^{2-}\right]$, we see that the only mathematical operations involved are multiplication and division. Therefore we can evaluate the overall uncertainty by using the rule for fractional uncertainties given above. First, we need to calculate the fractional uncertainty of each quantity.

$$\text{Fractional uncertainty on titre} = \pm \frac{0.07\,\text{cm}^3}{14.85\,\text{cm}^3} = \pm 0.004\,7$$

Fractional uncertainty on permanganate concentration

$$= \pm \frac{0.000\,05\,\text{mol dm}^{-3}}{0.005\,0\,\text{mol dm}^{-3}}$$

$$= \pm 0.010\,0$$

$$\text{Fractional uncertainty on ethanedioate volume} = \pm \frac{0.1\,\text{cm}^3}{20.0\,\text{cm}^3} = \pm 0.005$$

The stoichiometric ratio $\dfrac{5}{2}$ is an exact number so its associated uncertainty is zero. Leaving it in this form avoids any requirement to consider the appropriate number of significant figures in its decimal equivalent. The overall fractional uncertainty is then

$$\sqrt{0.004\,7^2 + 0.010\,0^2 + 0.005^2} = \sqrt{0.000\,147} = 0.012\,1$$

We multiply this by our concentration value to give the absolute uncertainty

$$0.0121 \times 9.3 \times 10^{-3} \, \text{mol dm}^{-3} = 0.11 \times 10^{-3} \, \text{mol dm}^{-3}$$

The final value for the concentration is then given as $(9.3 \pm 0.1) \times 10^{-3}$ mol dm^{-3}. When scientific notation is involved as here, it is usually clearer to quote the uncertainty as a number to the same power of 10 as the value to which it refers, even if it means that the uncertainty is not given as a number between 1 and 10.

Worked Example 2.8

The concentrations of hydrogen, iodine and hydrogen iodide in the gaseous equilibrium

$$H_2 + I_2 \rightleftharpoons 2HI$$

are related by the equilibrium constant K, so that if all other quantities are known the hydrogen concentration can be determined from the equation

$$\left[H_2 \right] = \frac{\left[HI \right]^2}{K \left[I_2 \right]}$$

If the equilibrium constant K has been determined as 46.0 ± 0.2, calculate the concentration of hydrogen, with its associated uncertainty, when $[HI] = (17.2 \pm 0.2) \times 10^{-3}$ mol dm^{-3} and $[I_2] = (2.9 \pm 0.1) \times 10^{-3}$ mol dm^{-3}.

CHEMICAL BACKGROUND

This is an example of a chemical equilibrium which has been studied extensively. Although we have dealt with this in terms of concentrations, it is also possible to do so in terms of the partial pressures of each component. No matter what the starting concentrations or pressures, at equilibrium, the relationship given will always be obeyed at a given temperature; at a different temperature, we would have to use a different value for the equilibrium constant K. One consequence of this relationship is that it does not matter whether we start with a mixture of hydrogen or iodine, or by allowing hydrogen iodide to decompose. The same equilibrium will be reached in each case.

One practical way in which this equilibrium could be studied would be to heat hydrogen iodide in a bulb of known volume as shown in Figure 2.5. Once equilibrium has been reached, this could then be cooled rapidly to room temperature to stop or 'quench' the reaction. The concentration of iodine can then be determined by titration with sodium thiosulphate.

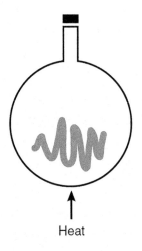

Heat

FIGURE 2.5 Measuring the equilibrium constant for the dissociation of hydrogen iodide.

Solution to Worked Example 2.8

Using the equation given to evaluate the concentration of hydrogen gives us

$$[H_2] = \frac{(17.2 \times 10^{-3} \, mol\,dm^{-3})^2}{46.0 \times 2.9 \times 10^{-3} \, mol\,dm^{-3}}$$

which can be evaluated using a calculator to give the value of 2.218×10^{-3}. Since we are going to evaluate an uncertainty limit from the information given there is no point at this stage worrying too much about the number of significant figures to quote. It is best to err on the side of caution, however, to make sure that we include enough significant figures to allow rounding at a later stage.

Notice that, since only multiplication and division are involved in the equation given, we need to combine the fractional uncertainty on each quantity. Since the equation contains $[HI]^2$ rather than $[HI]$, the first thing we will do is to evaluate the fractional uncertainty on $[HI]^2$, which is

$$\frac{0.2}{17.2} = 1.2 \times 10^{-2}$$

Since the concentrations and their associated uncertainties are multiplied by the same powers of 10, these can be ignored when we are dealing with fractional uncertainties. Using the appropriate equation from Section 2.4.2.2, we have

Fractional error on $[HI]^2 = \sqrt{2} \times$ fractional error on $[HI]$

$$= \sqrt{2} \times 1.2 \times 10^{-2}$$

$$= 1.7 \times 10^{-2}$$

The remaining fractional uncertainties are straightforward to calculate:

Fractional uncertainty on $K = \dfrac{0.2}{46.0} = 4.3 \times 10^{-3}$

Fractional uncertainty on $[I_2] = \dfrac{0.1}{2.9} = 3.4 \times 10^{-2}$

We now combine these three fractional uncertainties using the rule for products and quotients:

Fractional uncertainty on $[H_2] = \sqrt{(1.7 \times 10^{-2})^2 + (4.3 \times 10^{-3})^2 + (3.4 \times 10^{-2})^2}$

$$= 0.038\,2$$

To obtain the absolute uncertainty on H_2 we need to multiply this fractional uncertainty by the value of $[H_2]$ itself.

Absolute uncertainty on $[H_2] = 0.038\ 2 \times 2.218 \times 10^{-3}\,\text{mol dm}^{-3}$

$$= 0.084\ 727\ 6 \times 10^{-3}\,\text{mol dm}^{-3}$$

This could also be written as $8.47 \times 10^{-5}\,\text{mol dm}^{-3}$, but if we leave it in the form above we can now write our final expression for the hydrogen concentration as

$$[H_2] = (2.22 \pm 0.08) \times 10^{-3}\,\text{mol dm}^{-3}$$

where the uncertainty is expressed as a single digit in the appropriate decimal place to reflect its calculated value.

2.4.3 STATISTICAL TREATMENT OF UNCERTAINTIES

Statistics is a discipline in itself, so we will do no more than look at the basics here to enable us to assign reasonable uncertainty limits. As noted earlier, the statistical treatment is only of use when we have repeated measurements of the same quantity. For our purposes, it is only important to understand a few basic definitions of statistical quantities and to be able to apply them to sets of data.

To illustrate these definitions, consider the small data set

1.2	1.4	1.2	1.5	1.3

which contains five numbers. One way of representing these is by using the notation x_i, where i is an integer so that x_1 is the first number (1.2), x_2 is the second number (1.4), and so on.

The **arithmetic mean** is obtained by adding all the values of the quantity together and then dividing by the number of measurements taken. It can be represented as

$$\bar{x} = \frac{\Sigma x_i}{n}$$

where n is the number of numbers x_i in the data being considered.

Do not be put off by the use of some strange-looking mathematical symbols here. This is a good place to introduce some terminology as the concepts are still simple, and it will make life easier with some of the other quantities. The symbol Σ (a capital Greek sigma) simply means 'add up'. So if we add the five numbers given above we get 6.6, which is equal to Σx_i. We then divide this by n, the number of values which, in this case, is 5. This gives us a value for the arithmetic mean of

$$\frac{6.6}{5} = 1.3$$

using our previous rules for significant figures (Section 2.2.2) and noting that the integer 5 is taken as an exact number.

The **median** is obtained by arranging all the values obtained in ascending (or descending) order; if there are an odd number of values, the median is the middle one; otherwise, it is the arithmetic mean of the two middle values. So, arranging our data set in ascending order (lowest first) gives us

$$1.2 \quad 1.2 \quad 1.3 \quad 1.4 \quad 1.5$$

As we have an odd slumber of values, the central one is the third, and so the value of the median is 1.3.

The **mode** is the most frequently occurring value. Looking at our example, only 1.2 appears more than once, so this is the mode in this case.

The arithmetic mean, median and mode are different forms of average. It is also useful to be able to determine the extent to which a set of data is spread. The easiest way of doing this is to measure the **range**, which is the difference between the maximum and minimum values. In the data set above, this will be

$$1.5 - 1.2 = 0.3$$

A major limitation with the range is that if there are any unusually high or low values this will distort its value. To take account of this more sophisticated measures of spread have been developed.

Before considering these we need to make a distinction between our sample of measurements and the population from which they are derived. In general, we will not be able to measure all the possible values from the population of interest, but have to rely on a smaller number of measurements which constitute our sample. The mean of the population is generally denoted by the Greek letter μ, while that of a sample is represented by $\bar{x}$.

Hopefully, it will be intuitive that a reasonable way to measure spread is to consider the average difference between each individual value and the mean. However, if we just added up the differences we would always obtain a value of 0 as some differences would be positive and some negative. For example, in our current data set, we would have

$$1.2 - 1.32 = -0.12$$

$$1.4 - 1.32 = 0.08$$

$$1.2 - 1.32 = -0.12$$

$$1.5 - 1.32 = 0.18$$

$$1.3 - 1.32 = -0.02$$

The sum of these is zero. Note that we have carried more figures than justified in the intermediate calculations to avoid rounding errors.

To address this we actually take the average of the squares of the differences from the mean. For the population, this gives us a quantity called the variance, defined by the equation

$$\sigma^2 = \frac{\Sigma(x_i - \mu)^2}{n}$$

The equation for the variance of the sample is slightly different in that instead of dividing by n we divide by $n-1$:

$$s^2 = \frac{\Sigma(x_i - \bar{x})^2}{n-1}$$

The division by $n-1$ rather than n allows for the fact that the variance calculated for the sample is likely to underestimate that of the population; as $n-1$ is smaller than n this will have the effect of increasing s^2. The difference is less apparent for large values of n. This is not surprising if we consider that as n increases it approaches the size of the whole population.

You may have noticed that the symbols used for the variance of the population σ^2, and variance of the sample s^2 both involve a squared quantity. Because we are taking the square of the difference between each value and the mean, the variance will have units which are the square of those of the original readings x_i. It therefore is more useful to take the square roots of μ^2 and s^2; μ and s are known as the standard deviations of the population and sample respectively and have the same units as the quantity being measured.

It can be shown that 95% of all values would be expected to lie within two standard deviations of the mean, so by convention two standard deviations can be taken as being a reasonable uncertainty estimate for an individual reading.

If we take several samples, each of size n, from our population then it is likely that a different mean will be obtained for each one. We can then determine the standard deviation of the mean, which is commonly known as the standard error of the mean and defined as

$$\frac{\sigma}{\sqrt{n}}$$

There may be times in chemistry when our sample is the same as the population. This would be the case when we have prepared a relatively small number of batches of a compound in the laboratory. Each can then be analysed to give the quantity required. This is not the case if we wished to analyse the composition of a naturally occurring substance, where sampling is the only possible strategy.

Worked Example 2.9

Ten determinations of the dipole moment of hydrogen chloride gas gave the following values (in units of D):

1.048	1.047	1.053	1.048	1.051
1.053	1.045	1.051	1.047	1.047

Calculate (a) the arithmetic mean, (b) the median, (c) the mode, (d) the variance and (e) the standard deviation of these values.

CHEMICAL BACKGROUND

When placed in an electric field, all molecules have an induced dipole moment, since the centres of gravity of positive and negative charges do not coincide. Polar molecules also have a permanent dipole moment (Figure 2.6) which can be determined from measurements of the dielectric constant, also called the relative permittivity. In the case of hydrogen chloride, this can be done using a specially constructed gas capacitance cell.

Solution to Worked Example 2.9

a. The sum of the values is 10.49 D, so the arithmetic mean is

$$\frac{10.49\,D}{10} = 1.049\,D$$

Notice that by using the rules of significant figures (with 10 being an exact integer) we keep the same number of decimal places as in the original data.

b. Rearranging the data in ascending order gives

1.045	1.047	1.047	1.047	1.048

1.048	1.051	1.051	1.053	1.053

Since there are an even number of data we need to look at the fifth and sixth values; these are actually the same so their arithmetic mean and the median of the data is 1.048 D.

c. By inspection, the value of 1.047 D appears three times and as this is the most frequent value this is the mode.

d. To calculate the variance, we need to calculate the sum of the squares of the deviation from the mean for each value which occurs and then divide by $n-1$:

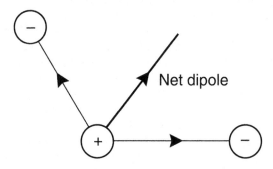

FIGURE 2.6   Net dipole resulting from two polar bonds.

$$\sigma^2 = \left(\frac{1}{10-1}\right) \times \begin{bmatrix} 1 \times (1.045 - 1.049)^2 \\ +3 \times (1.047 - 1.049)^2 \\ +2 \times (1.048 - 1.049)^2 \\ +2 \times (1.051 - 1.049)^2 \\ +2 \times (1.053 - 1.049)^2 \end{bmatrix} D^2$$

$$\sigma^2 = \left(\frac{1}{9}\right)[1.6 + 1.2 + 0.2 + 0.8 + 3.2] \times 10^{-5} \, D^2$$

$$= 7.8 \times 10^{-6} \, D^2$$

Note the use of standard form notation (see Section 1.7) in this calculation to make the manipulation of the numbers more manageable.

e. The standard deviation is the square root of this value, which is 0.003 D. It is usual to quote the standard deviation to the same number of decimal places as the original data.

2.4.3.1 STATISTICS USING A CALCULATOR

It is difficult to provide detailed instructions for performing statistical calculations on a calculator due to the large variation in the keystrokes required. However, I will try to indicate the keys that you need to look for. For detailed instructions see the manual for your own machine. There are likely to be three or four steps that you need to take.

1. On some calculators there is the facility to clear the statistics memory. This is likely to involve a combination of the keys labelled **SHIFT** and **MODE**, along with perhaps the input of a numerical option. This may be labelled on the screen with text such as **Scl** (for **S**tatistics **cl**ear).
2. On some calculators this will be the first step. It is to select the statistics mode of calculation, and will often involve the selection of the **MODE** key along with a numerical input which may be labelled on the screen with text such as **SD**.
3. The next step is to enter the data using the numerical keys. Each value will need to be followed by pressing a key such as those labelled **M+** and/or **DT**. It may also be necessary to indicate the end of data input by pressing a key such as **AC**.
4. The final step is to tell the calculator which calculation you wish to perform. Some calculators require pressing a key which causes a numerical menu to appear on the screen while others may have a dedicated key such as **S-VAR** which produces the same type of display when pressed in conjunction with the **SHIFT** key.

The resulting menus are likely to contain quantities such as $\bar{x}$, $x\sigma_n$ and $x\sigma_{n-1}$. These denote the mean, the standard deviation of the population and the standard deviation of the sample respectively, as discussed in the previous section.

Worked Example 2.10

The following determinations of the interatomic distance in hydrogen chloride were made (values in Å):

1.279	1.277	1.277	1.281
1.278	1.283	1.281	1.280

Use statistical methods to obtain an estimate of the H-Cl bond length with its uncertainty.

CHEMICAL BACKGROUND

The H-CI bond length can be determined by measuring the vibrational-rotational spectrum of hydrogen chloride gas, a part of which is shown in Figure 2.7. This gives the value of the rotational constant B, which is defined by

$$B = \frac{h}{8\pi^2 I}$$

where h is Planck's constant and I is the moment of inertia defined by

$$I = \mu r^2$$

where μ is the reduced mass and r the is the internuclear distance. The reduced mass of two individual masses m_1 and m_2 is defined as

$$\mu = \frac{m_1 m_2}{m_1 + m_2}$$

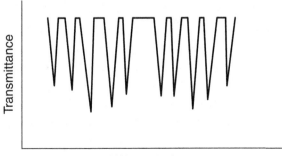

FIGURE 2.7   The vibration-rotation spectrum of hydrogen chloride gas.

The angstrom (Å) is a non-SI unit which still finds much favour among chemists working on structural problems. It is defined as 10^{-10}m, and consequently a single bond between two carbon atoms has a length of 1.54 Å. The use of numbers around 1 and 2 is preferred to the fractional nm (e.g. C-C length 0.154 nm) or larger pm (C-C length 154 pm) in some situations.

Solution to Worked Example 2.10

We need to calculate the standard deviation of these data. The quickest way to do this is using a calculator, but here the manual working is also shown. You may wish to replicate this calculation by working through the individual steps as well as using the statistical calculator functions described above. We begin by calculating the arithmetic mean.

$$\text{Sum of data} = 10.236\,\text{Å}$$

$$\text{Arithmetic mean} = \frac{10.236}{8} = 1.280\,\text{Å}$$

which is given to 3 decimal places as was the original data.

The next stage is to obtain the variance.

$$\sigma^2 = \left(\frac{1}{8-1}\right) \times [2 \times (1.277 - 1.280)^2 + 1 \times (1.278 - 1.280)^2$$

$$+ 1 \times (1.279 - 1.280)^2 + 1 \times (1.280 - 1.280)^2$$

$$+ 2 \times (1.281 - 1.280)^2 + 1 \times (1.283 - 1.280)^2]$$

$$= \left(\frac{1}{7}\right) \times [1.8 + 0.4 + 0.1 + 0.0 + 0.2 + 0.9] \times 10^{-6}\,\text{Å}^2$$

$$= 4.86 \times 10^{-6}\,\text{Å}^2$$

This calculation has used the equation for the variation of a sample; here we have a sample of H-Cl bond lengths taken from a much larger population.

The standard deviation is the square root of this, 0.002 2 Å. It then follows that the standard error will be

$$\frac{\sigma}{n} = \frac{0.002\,2\,\text{Å}}{\sqrt{8}}$$

$$= \frac{0.002\,2\,\text{Å}}{2.828}$$

$$= 8 \times 10^{-4}\,\text{Å}$$

If we take twice the standard error of 1.6×10^{-3}Å as a reasonable uncertainty estimate, we have the H-Cl bond length as 1.280 ± 0.002 Å. It often makes no sense to quote the uncertainty to any more figures than in the final decimal place of the original data.

2.4.3.2 STATISTICS USING A SPREADSHEET

While the formula for the calculation of standard deviation is straightforward to apply, this can be rather tedious if we have a large number of data, even if a calculator is used. In such cases, it makes sense to take advantage of one of the great features of computers, namely that they are able to rapidly perform repetitive calculations. The most straightforward way to do this is to make use of one of the commonly available spreadsheet programs.

The commonly used spreadsheets have functions for calculating the standard deviation of both the sample and the population, the difference between which was explained in Section 2.4.3.

In Excel, the relevant functions are STED.S and STDEV.P respectively, while in Google Sheets the equivalent functions are STDEV and STDEVP.

Worked Example 2.11

The value for the C=O stretching frequency in a number of compounds belonging to several classes of carbonyl compounds is summarized in the table below, in terms of wavenumber measured in cm^{-1}. Determine the standard deviation for this set of data using the most efficient method.

Aldehyde	1722	1722	1739	1727	1728	1731	1727	1721	1738	1732
Ketone	1721	1708	1715	1713	1710	1719	1720	1710	1712	1708
Carboxylic acid	1721	1719	1716	1702	1717	1707	1711	1711	1719	1703
Ester	1742	1737	1749	1748	1739	1745	1749	1737	1736	1746
Carboxylate	1555	1601	1594	1592	1607	1559	1562	1568	1577	1580

CHEMICAL BACKGROUND

Infrared spectroscopy is a very powerful technique for identifying compounds. This is due to the fact that specified functional groups have characteristic vibrational frequencies in specified environments. This can be seen in the table above, where the C=O stretching frequency is between 1720 and $1740\,cm^{-1}$ in aldehydes, 1705 and $1725\,cm^{-1}$ in ketones, 1700 and $1725\,cm^{-1}$ in carboxylic acids, 1735 and $1750\,cm^{-1}$ in esters and 1550 and $1610\,cm^{-1}$ in carboxylate ions.

Solution to Worked Example 2.11

Given the relatively large number of data, the most efficient method will be to use a spreadsheet. It would take a long time to input this on a calculator, and there is a high possibility of making one or more mistakes while doing so.

Figure 2.8 shows the way of solving this using Microsoft Excel. The labels for the data are input to cells A2–A6, although they are not actually needed in order to perform the calculation. The numerical data appears in cells B2–K6, which can be referenced in the

FIGURE 2.8 Microsoft Excel spreadsheet showing data for Worked Example 2.11.

FIGURE 2.9 Google Sheets spreadsheet showing data for Worked Example 2.11.

spreadsheet as B2:K6. Cell B8 is set up to receive the value of the standard deviation; this is highlighted in the figure to display the formula =STEDV.S(B2:K6). Note the use of the=symbol to indicate that a formula is being input. The function STDEV.S has been selected to calculate the standard deviation of the sample. This gives a value for the standard deviation of $60 \, \text{cm}^{-1}$.

We also need to be aware that spreadsheet functions will only operate on numbers, so we need to take care with the units of any resulting values. In this case, we know that the standard deviation will have the same units as the original data so it is straightforward to quote the units as cm^{-1}. Note also that the format of cell B8 has been set to contain a number with zero decimal points. This ensures that the standard deviation is quoted to the same precision as the original data. This feature can be accessed by the Number tab on the home ribbon.

The calculation is performed in a very similar way in Google Sheets, as shown in Figure 2.9. Notice that cell B8 now contains the formula =STDEV(B2:K6). The biggest difference from Excel is in formatting B8 correctly; this requires selection of the sequence Format, Number, More Formats, Choose number format, 0.

EXERCISES

1. Write the following numbers to 2 significant figures:
   (a) 348   (b) 0.001 2   (c) 0.543 7   (d) 12.65   (e) 4286

2. Write the following numbers to 2 decimal places:
   (a) 13.847   (b) 0.254 9   (b) 99.543 21   (c) 0.007 54   (d) 1200

3. Write the following numbers to 3 significant figures:
   (a) 78 521   (b) 0.006 754 3   (c) 12.005 485   (d) 80 006
   (e) 0.318 75

4. Write the following numbers to 1 decimal place:
   (a) 3.098   (b) 12.76   (c) 0.000 62   (d) 1.049   (e) 0.995

5. Give the results of the following calculations to the appropriate number of figures:
   (a) 12.36+3.2   (b) 31.8 − 12.72   (c) 1.9×1.235

   (d) $\dfrac{4.673}{2.1}$ (e) $\dfrac{(34.2-1.75)}{2.34}$ (f) $(3.1\times2.71)+4.92$

6. Give the results of the following calculations to the appropriate number of significant figures:

   (a) 2×32.74   (b) 3×12.01×32.1   (c) $\left(\dfrac{5}{2}\right)\times\left(\dfrac{3.07}{6.18}\right)$

   (d) $\left(\dfrac{1}{3}\right)\times(2.96+3.8)$

7. Give the results of the following calculations to the appropriate number of figures:

   (a) $1.05 + 2.3 \times 4.08$ (b) $(2.49 - 1.06) \times 3.1$ (c) $4.989 - \dfrac{4.08}{1.249}$

   (d) $\dfrac{2.61\times7.02}{1.36+0.784}$

8. Give the results of the following calculations to the appropriate number of figures:

   (a) $9.714 - 2.62^2$ (b) $\dfrac{3.19 - 2.584}{1.2^3}$ (c) $\dfrac{(2.195 + 3.42)^2}{8.714\,8}$

   (d) $\dfrac{(1.48 + 2.976)^2}{(3.065 - 1.23)^3}$

9. A number is quoted as 32.6±0.4. What is the (a) absolute uncertainty, (b) the fractional uncertainty and (c) the percentage uncertainty?

10. The sides of a cuboid are measured as 21.0±0.1 cm, 16.0±0.1 cm and 32.0±0.2 cm. What is (a) the maximum possible absolute uncertainty, (b) the maximum possible fractional uncertainty and (c) the maximum possible percentage uncertainty in the volume of the cuboid?

11. Repeat question 8 but this time calculate the maximum probable uncertainties in each case.

12. Three variables are measured as $x=3.07\pm0.05$, $y=4.29\pm0.08$ and $z=2.37\pm0.10$. Determine the maximum probable uncertainty in

   (a) $x+y-z$ (b) $xy+z$ (c) $x+\dfrac{y}{z}$ (d) $\dfrac{x}{y}+xz$

13. Four variables are measured as $a = 9.143 \pm 0.008$, $b = 6.413 \pm 0.007$, $c = 2.985 \pm 0.05$ and $d = 8.764 \pm 0.006$. Determine the maximum probable uncertainty in

   (a) $\dfrac{d}{c} - \dfrac{a}{b}$ (b) $a - b + \dfrac{c}{d}$ (c) $\dfrac{a}{b} + (d - c)^2$ (d) $\dfrac{(a-b)^3}{(c+d)^2}$

14. A car covered a distance of 9.25 ± 0.05 km in 295 ± 1 second. Use the formula

$$\text{speed} = \frac{\text{distance}}{\text{time}}$$

   to calculate the average speed of the car and its associated maximum probable uncertainty.

15. Calculate (a) the arithmetic mean, (b) the median and (c) the mode of the following numbers:

10.65	10.62	10.58	10.67	10.66	10.61	10.61

16. Determine the standard deviation of the following numbers by calculating each intermediate step:

23.72	24.01	23.86	24.09	23.68	23.99

17. Determine the standard error of the following numbers by calculating each intermediate step:

98.54	100.06	99.42	98.95	101.03	100.01	98.95	99.51

18. Use the statistical function on a calculator to determine the standard deviation of the following numbers:

3.79	3.61	3.84	3.76	3.72	3.82	3.66	3.73	3.65	3.75
3.83	3.71	3.62	3.65	3.66	3.83	3.62	3.81	3.85	3.74

19. Use the statistical function on a calculator to determine the standard deviation, and hence the standard error, of the following numbers:

72.1	74.8	73.7	74.1	73.2	74.3

20. Use a spreadsheet to determine the standard deviation of the following numbers:

19.45	19.42	19.51	19.55	19.53	19.45	9.45	19.56	19.52	19.49
19.44	19.50	19.50	19.56	19.47	19.44	19.49	19.48	19.44	19.47
19.46	19.43	19.49	19.51	19.47	19.49	19.55	19.54	19.50	19.49
19.55	19.53	19.47	19.48	19.52	19.55	19.53	19.47	19.45	19.44

PROBLEMS

1. Write the values of each of these quantities to (a) 3 significant figures and (b) 2 decimal places:
   a. a bond length of 1.542 Å
   b. a temperature of 25.01°C
   c. an enthalpy change of $-432.876 \, kJ \, mol^{-1}$
   d. the relative atomic mass of oxygen, 15.999 4
   e. the gas constant, $8.314 \, J \, K^{-1} mol^{-1}$

2. Give the results of the following calculations to the appropriate number of figures:
   a. $density = \dfrac{mass}{volume} = \dfrac{17.098 \, g}{6.2 \, cm^3}$
   b. mass of sample $= 20.145 \; 2 \, g - 5.12 \, g$
   c. internal energy $\Delta U = Q + W = 321.4 \, kJ \, mol^{-1} - 162 \, kJ \, mol^{-1}$
   d. $pV = nRT = 3.2 \, mol \times 8.314 \, J \, K^{-1} mol^{-1} \times 298 \; K$
   e. equilibrium constant $K_c = \dfrac{0.106 \, mol \, dm^{-3} \times 0.098 \, mol \, dm^{-3}}{0.250 \, mol \, dm^{-3}}$

3. Give the results of the following calculations to an appropriate number of figures:
   a. $\Delta G = -246 \, kJ \, mol^{-1} - 298 \; K \times 0.124 \, kJ \, K^{-1} mol^{-1}$
   b. $C_p = 19.875 \, J \, K^{-1} mol^{-1} + 5.021 \times 10^{-2} J \, K^{-2} mol^{-1} \times 305.4 \; K$
   c. $\dfrac{1}{c} = \dfrac{1}{7.50 \times 10^{-3} \, mol \, dm^{-3}} + 0.059 \, dm^3 \, mol^{-1} \, min^{-1} \times 2.36 \, min$

4. The virial equation can be expressed as

$$Z = 1 + \frac{Bp}{RT}$$

   where p represents pressure, T the absolute temperature, and R the gas constant $8.314 \, J \, K^{-1} mol^{-1}$. Calculate Z for helium for which $B = 11.8 \times 10^{-6} m^3 mol^{-1}$ at 298.15 K. Take the pressure as 250 kPa.

5. The overall entropy change for any process is equal to the sum of the entropy changes for the individual steps involved. In one experiment, these changes were measured as $3.90 \pm 0.05 \, J \, K^{-1} mol^{-1}$, $38.4 \pm 0.1 \, J \, K^{-1} mol^{-1}$ and $5.05 \pm 0.05 \, J \, K^{-1} mol^{-1}$. Determine the maximum possible uncertainty and the maximum probable uncertainty in the sum of these three entropy changes.

6. The volume V of an ideal gas can be found by rearranging the ideal gas equation, and is given by

$$V = \frac{nRT}{p}$$

where n is the amount of gas, R is the gas constant (8.314 J $K^{-1}mol^{-1}$), T is the absolute temperature and p is the pressure. Calculate the volume, with its associated uncertainty, for 2.35 ± 0.1 mol of gas at a temperature of 298.4 ± 0.1 K and a pressure of 1.033 ± 0.005 kPa.

7. Concentration may be expressed in terms of the mass percentage of solute, which is the percentage by mass of solute contained in a solution. A solution of sodium chloride was prepared by dissolving 2.67 ± 0.04 g of the salt in 84.3 g of water. Determine the mass percentage of solute in this case, together with the maximum probable uncertainty.

8. The wavelength of one of the D-lines in the sodium spectrum was measured (in nm) as follows:

589.595 9	589.595 5	589.596 2
589.597 4	589.608 1	589.593 0
589.592 2	589.590 8	589.585 9
589.575 6	589.578 1	589.600 2

Determine the mean and median of this data.

9. Calculate the arithmetic mean, median, mode, variance and standard deviation of the following values of the ionization energy of sodium (in kJ mol^{-1}):

495	497	493	493	496	492

10. Use a spreadsheet to determine the standard deviation of the following Cu-N bond lengths (in Å) in a series of compounds as determined by X-ray crystallography:

2.035 2.038 2.047 2.052 2.051 2.039 2.040 2.052 2.036 2.041
2.039 2.038 2.050 2.049 2.053 2.048 2.041 2.039 2.040 2.037
2.051 2.053 2.038 2.053 2.038 2.052 2.039 2.036 2.051 2.036

11. Calculate the standard error on the following measurements of the molar volume of an ideal gas (in dm^3):

22.406	22.412	22.410	22.408	22.409	22.414

12. The following values (in kJ mol^{-1}) were obtained for the enthalpy of combustion of methane according to the equation

$$CH_{4(g)}+2\,O_{2(g)} \rightarrow CO_{2(g)}+2\,H_2O_{(l)}$$

−892.4	−891.6	−891.7	−890.1	−889.8
−890.9	−890.9	−888.9	−889.9	−891.4

Determine the 95% confidence limit for this data.

Thermodynamics

3.1 FRACTIONS AND INDICES IN THE EQUILIBRIUM CONSTANT

Thermodynamics is one of the branches of physical chemistry which students often find most confusing. It involves many different concepts, each of which seems to require a different mathematical approach and a different terminology. However, it is also a very important subject, whose ideas underpin the whole of chemistry and without which a full appreciation of chemical concepts is not possible.

In this chapter, we will work through some of the main areas of thermodynamics and explore the mathematics required to deal with each of them. We will cover a number of different areas of mathematics, including what may be the first experience you have had of calculus. The terminology used can sometimes be offputting, but each new symbol will be explained as we meet it. You will find the maths easier if you can remember the meanings of the symbols introduced, so you may wish to make a careful note of these as we proceed.

The equilibrium constant is a fundamental quantity for any chemical reaction in which reactants and products are in equilibrium. Equilibrium constants are important industrially because they allow the conditions which optimize the production of the desired products to be determined.

For the general chemical equilibrium

$$aA + bB \rightleftharpoons cC + dD$$

in which A, B, C and D represent chemical symbols and a, b, c and d are stoichiometric coefficients, the equilibrium constant K_c in terms of concentrations is defined by

$$K_c = \frac{[C]^c[D]^d}{[A]^a[B]^b}$$

where the square brackets represent the concentration of each chemical species. We can write down an expression for the equilibrium constant as long as we know the equation for the equilibrium reaction.

DOI: 10.1201/9781003043218-3

Worked Example 3.1

A volume of 1.0 dm³ of a gaseous mixture contains 0.30 mol of CO, 0.10 mol of H_2, 0.020 mol of H_2O and 0.059 mol of CH_4 at a temperature of 1200 K. What is the equilibrium constant for the equilibrium reaction

$$CO_{(g)} + 3H_{2(g)} \rightleftharpoons CH_{4(g)} + H_2O_{(g)}$$

at this temperature?

CHEMICAL BACKGROUND

K_c specifies the equilibrium constant in terms of concentrations. A second equilibrium constant K_p is specified in terms of pressure. While we might expect to calculate K_p for a gaseous reaction such as this, it is perfectly acceptable to calculate K_c if we know the amounts of each species present and the total volume.

Solution to Worked Example 3.1

We can immediately write down an expression for K_c, which is

$$K_c = \frac{[CH_4][H_2O]}{[CO][H_2]^3}$$

Since the stoichiometric coefficients are 1 apart from H_2 the indices for these concentrations are also 1 so do not have to be shown explicitly. The total volume of the mixture is 1.0 dm³ so the amounts specified in the question in mol are numerically equal to their concentrations in mol dm⁻³. The equilibrium constant is therefore

$$K_c = \frac{0.059 \text{ mol dm}^{-3} \times 0.020 \text{ mol dm}^{-3}}{0.30 \text{ mol dm}^{-3} \times (0.10 \text{ mol dm}^{-3})^3} = 3.9 (\text{mol dm}^{-3})^{1+1-(1+3)}$$

$$= 3.9 \, (\text{mol dm}^{-3})^{-2}$$

$$= 3.9 \, \text{mol}^{-2} \, \text{dm}^6$$

$$= 3.9 \, \text{dm}^6 \, \text{mol}^{-2}$$

Note that the units have been included in the calculation and that the rules of indices from Section 1.6 have been applied to allow us to combine those units.

3.2 BOND ENTHALPIES

The idea that each type of bond has associated energy (at constant pressure this is known as the bond enthalpy) gives us a straightforward way of estimating the enthalpy change for a given chemical reaction. If we assume that all chemical bonds in the reactants are broken, requiring positive enthalpy input, and that all chemical bonds in the products are formed, releasing negative enthalpy then the overall enthalpy change will be the sum of the individual enthalpy changes.

3.2.1 REARRANGING EQUATIONS

The mathematical background to this was discussed in Section 1.4 on page 9. If we have a relationship such as

$$w=x+y+z$$

and we wish to obtain an expression for y we subtract x and z from each side to give

$$w-x-z=x+y+z-x-z$$

or

$$y=w-x-z$$

Worked Example 3.2

The enthalpy change for the reaction

$$CH_{4(g)}+Cl_{2(g)} \rightarrow CH_3Cl_{(g)}+HCl_{(g)}$$

is −104 kJ mol⁻¹. If the bond enthalpy for C-H is 411 kJ mol⁻¹, for H-Cl is 428 kJ mol⁻¹, and for C-Cl is 327 kJ mol⁻¹, what is the bond enthalpy for Cl-Cl?

CHEMICAL BACKGROUND

This is one in a series of reactions which add chlorine to methane to give successively CH_2Cl_2, $CHCl_3$ and CCl_4.

Bond enthalpies only allow an estimate of the overall enthalpy change for a reaction because they are average values measured across a large range of compounds.

Solution to Worked Example 3.2

In the reactants, we need to break 4 C-H and 1 Cl-Cl bonds. The enthalpy required to do this is 4×411 kJ mol⁻¹+x, where x is the Cl-Cl bond enthalpy we wish to find. This is equal to 1644 kJ mol⁻¹+x.

In the products, we form 3 C-H, 1 C-Cl and 1 H-Cl bond. The total enthalpy released will thus be 3×411 kJ mol⁻¹+327 kJ mol⁻¹+ 428 kJ mol⁻¹, which is equal to 1988 kJ mol⁻¹; since this is enthalpy released we take the negative value.

The overall enthalpy change for the reaction is then

$$1644 \text{ kJ mol}^{-1}+x-1988 \text{ kJ mol}^{-1}=-104 \text{ kJ mol}^{-1}$$

To solve this equation for x we need to subtract 1644 kJ mol^{-1} and add 1988 kJ mol^{-1} to each side. This gives

$$1644 \text{ kJ mol}^{-1} + x - 1988 \text{ kJ mol}^{-1} - 1644 \text{ kJ mol}^{-1} + 1988 \text{ kJ mol}^{-1}$$
$$= -104 \text{ kJ mol}^{-1} - 1644 \text{ kJ mol}^{-1} + 1988 \text{ kJ mol}^{-1}$$

or

$$x = 240 \text{ kJ mol}^{-1}$$

3.3 THE BORN-HABER CYCLE

One of the fundamental ideas in thermodynamics is that the change in the value of certain variables is equal to the sum of other smaller changes which give the same overall result. For example, when a chemical reaction involving some kind of molecular rearrangement takes place, the change in the enthalpy is the same as if we had dissociated all the reactants into atoms and allowed them to recombine into the required products. This idea formed the basis of Worked Example 3.2 above. Of course, we know that chemical reactions do not really happen in this way, but it is a useful concept which allows us to calculate quantities which are difficult to measure experimentally. An analogy from everyday life is that if we climb a mountain, the height we reach is independent of the path we choose to take.

Some of the thermodynamic variables for which this is true are those called the enthalpy, entropy, Gibbs free energy and chemical potential; these are known as state functions. The two most common non-state functions are work and heat. You will meet all of these variables at some point in your study of physical chemistry.

The mathematical consequence of using state functions is that we can calculate the value of a change in a particular fraction by adding together other changes which would constitute a different route, even though the starting and end points are the same. While this is very useful to chemists in being able to calculate enthalpy changes for reactions where they cannot be measured, there is an important point to remember. We saw in Chapter 2 that any experimental measurement is subject to an uncertainty, and if we are adding together several experimental values we are likely to obtain an overall uncertainty which is much higher. It follows that an enthalpy change calculated in this way is likely to be less certain than one which has been measured directly.

3.3.1 COMBINING UNCERTAINTIES

The mathematical background to this has been discussed in Section 2.4.2 on page 38 where we saw how to determine the maximum probable uncertainty. We saw that if two quantities are measured as $X + \Delta X$ and $Y + \Delta X$, for sums and differences the absolute uncertainty is

$\sqrt{(\Delta X)^2 + (\Delta Y)^2}$, while for products and quotients the fractional uncer-

tainty is $\sqrt{\left(\dfrac{\Delta X}{X}\right)^2 + \left(\dfrac{\Delta Y}{Y}\right)^2}$.

Worked Example 3.3

In this example, we will use the symbol MX to denote a general-ized formula, where M represents a metal cation and X an anion. Examples would be CsCl and KF. These are 1:1 compounds where there are equal numbers of cations and anions. A 1:2 structure would have the generalized formula MX_2, examples being CaF_2 and MnS_2.

The standard lattice enthalpy $\Delta_{lattice}H^{\ominus}$ of a 1:1 ionic crystal MX is given by the equation

$$\Delta_{lattice}H^{\ominus} = -\Delta_f H^{\ominus} + \Delta_{sub}H^{\ominus} + \tfrac{1}{2}\Delta_{at}H^{\ominus} + \Delta_{ion}H^{\ominus} + \Delta_{ea}H^{\ominus}$$

where $\Delta_f H^{\ominus}$ is the standard enthalpy of formation of MX, $\Delta_{sub}H^{\ominus}$ is the standard enthalpy of sublimation of M, $\Delta_{at}H^{\ominus}$ is the standard enthalpy of atomization of X_2, $\Delta_{ion}H^{\ominus}$ is the ionization enthalpy of M and $\Delta_{ea}H^{\ominus}$ is the electron affinity of X.

Calculate the electron affinity of Cl, together with its associated uncertainty, using the following data:

Standard enthalpy of formation of $NaCl_{(s)} = -411 \pm 2$ kJ mol^{-1}

Enthalpy of sublimation of $Na = 107 \pm 1$ kJ mol^{-1}

Enthalpy of atomization of $Cl_{2(g)} = 244 \pm 1$ kJ mol^{-1}

Ionization enthalpy of $Na_{(g)} = 496 \pm 2$ kJ mol^{-1}

Standard lattice enthalpy of $NaCl = 787 \pm 3$ kJ mol^{-1}

CHEMICAL BACKGROUND

Sodium chloride exists as a regular cubic lattice, as shown in Figure 3.1. A consideration of the lattice enthalpy finds use when we are interested in the forces and energies within ionic crystals, as it is possible to compare the value with that obtained from theoretical calculations. In contrast to the problem here, we would normally calculate the value of the lattice enthalpy using experimental values of the other enthalpy changes; a direct experimental determination is not possible. The lattice enthalpy in this problem actually refers to the enthalpy change accompanying the reaction

$$NaCl_{(s)} \rightarrow Na^+_{(g)} + Cl^-_{(g)}$$

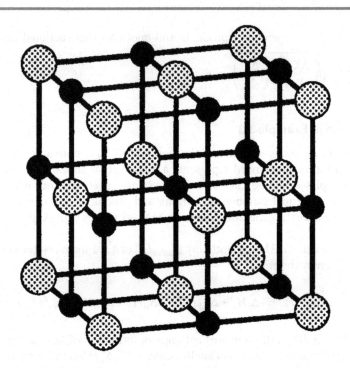

FIGURE 3.1 The crystal lattice of sodium chloride.

This contrasts with the enthalpy of formation, which is the enthalpy change for the reaction

$$Na_{(s)} + \tfrac{1}{2}Cl_{2(g)} \rightarrow NaCl_{(s)}$$

The other enthalpy changes required are the enthalpy of sublimation:

$$Na_{(s)} \rightarrow Na_{(g)}$$

and the enthalpy of ionization

$$Na_{(g)} \rightarrow Na^{+}_{(g)} + e^{-}$$

For anions, we can define the electron affinity, which in this case is the enthalpy change for the reaction

$$Cl^{-}_{(g)} \rightarrow Cl_{(g)} + e^{-}$$

This reaction is the reverse of what you might expect, and it is often written the other way round with accompanying phrases used to obtain the correct sign for the enthalpy change.

Solution to Worked Example 3.3

The calculation of $\Delta_{ea}H^{\ominus}$ is straightforward and involves rearranging the given equation to leave this term as the subject. This gives

$$\Delta_{ea}H^\ominus = \Delta_{lattice}H^\ominus + \Delta_f H^\ominus - \Delta_{sub}H^\ominus - \tfrac{1}{2}\Delta_{at}H^\ominus - \Delta_{ion}H^\ominus$$

$$= [787 - 411 - 107 - (0.5 \times 244) - 496] \text{ kJ mol}^{-1}$$

$$= -349 \text{ kJ mol}^{-1}$$

We can now turn our attention to determining the uncertainty in this quantity. Look first of all at the enthalpy of atomization of $Cl_{2(g)}$. This has been multiplied by 0.5, which we can take as being an exact constant with zero error. Therefore the fractional error on $0.5\Delta_{at}H^\ominus$ will be given by

$$\sqrt{\left(\frac{0}{0.5}\right)^2 + \left(\frac{1}{244}\right)^2} = \frac{1}{244}$$

the same as the fractional error on $\Delta_{at}H^\ominus$. The absolute uncertainty in this quantity will therefore be

$$\left(\frac{1}{244}\right) \times 0.5 \times 244 \text{ kJ mol}^{-1} = 0.5 \text{ kJ mol}^{-1}$$

Having done this, we have a set of uncertainties on five values which are combined by means of addition and subtraction. The absolute error is then given by

$$\sqrt{(3)^2 + (2)^2 + (1)^2 + (0.5)^2 + (2)^2} \text{ kJ mol}^{-1} = \sqrt{18.25} \text{ kJ mol}^{-1}$$

$$= 4.3 \text{ kJ mol}^{-1}$$

The value of $\Delta_{ea}H^\ominus$ is now -349 ± 4 kJ mol^{-1}.

Notice that the overall uncertainty is, inevitably, greater than any of the individual uncertainties used in the calculation. For this reason, as discussed earlier, a direct experimental measurement may be preferable to a calculated value such as this whenever possible.

3.4 HEAT CAPACITY

The branch of thermodynamics concerned with heat changes is thermo-chemistry, and the principal property which governs the thermal behaviour of a substance is the heat capacity, which can be defined at a simple level as the heat required to raise the temperature of the substance by 1 K. More precisely, we can identify two values of the heat capacity according to the conditions under which any change is brought about. They are defined as

$$C_p = \text{heat capacity at constant pressure}$$

and

$$C_v = \text{heat capacity at constant volume}$$

Frequently we are concerned with the molar heat capacity which then allows us to take into account of the amount of substance present.

3.4.1 EXPANSION OF BRACKETS

Mathematical expressions can be made more concise by the use of brackets. This is generally useful for giving general expressions, but there are times when it is desirable to expand such an expression into its individual terms. For example, if we have an expression such as

$$a(b+c)$$

this is expanded by multiplying the quantity outside the bracket (in this case, a) with each term inside the bracket.

$$a(b+c)=ab+ac$$

If two brackets are multiplied together, such as

$$(a+b)(c+d)$$

then each term in the first bracket needs to be multiplied with each term in the second.

$$(a+b)(c+d)=a(c+d)+b(c+d)$$

Multiplying out again gives

$$(a+b)(c+d)=a(c+d)+b(c+d)=ac+ad+bc+bd$$

It may, of course, be possible to simplify the resulting expression in many cases.

Worked Example 3.4

The value of the enthalpy change ΔH is given in terms of the heat capacity C_p and the temperature change ΔT as

$$\Delta H = C_p \Delta T$$

Write an expression for the absolute enthalpy H_2 at temperature T_2 in terms of the absolute enthalpy H_1 at temperature T_1 and the heat capacity.

CHEMICAL BACKGROUND

This relationship arises directly from the definition of enthalpy, which is the quantity of heat supplied to a system at constant pressure. It can be calculated from the formula

$$H = U + pV$$

where U is the internal energy, p is the pressure and V is the volume.

The expression given in the problem is only valid when C_p can be considered to be independent of temperature. In Worked Example 3.5 we will see an example of C_p depending on temperature.

Solution to Worked Example 3.4

We have already met the use of the Δ notation when discussing enthalpy changes in Worked Example 3.3. While a quantity such as an enthalpy H does have absolute values, in thermodynamics we are frequently more interested in changes to these values. We use the notation ΔH to denote such an enthalpy change, where this is defined as

$$\Delta H = H_2 - H_1$$

The convention is always to subtract the initial value (in this case H_1) from the final value (H_2). Similarly, the change in temperature is given by

$$\Delta T = T_2 - T_1$$

and so the expression becomes

$$H_2 - H_1 = C_p(T_2 - T_1)$$

We now expand the right-hand side of this equation by multiplying the terms in the bracket by C_p to give

$$H_2 - H_1 = C_p T_2 - C_p T_1$$

We were asked to obtain an expression for H_2, so we rearrange the equation, making H_2 the subject, by adding the term H_1 to both sides.

$$H_2 = H_1 + C_p T_2 - C_p T_1$$

This is a simple problem, yet being able to perform manipulations like this is vital to a mastery of thermodynamics.

3.4.2 POLYNOMIAL EXPRESSIONS

You may have seen expressions like

$$5x^3 + 2x^2 + 3x + 10$$

which is an example of a **polynomial expression** in x. The characteristic of a polynomial is that it contains a series of terms consisting of the variable (in this case x) raised to an integral power. The highest value of these powers is called the **order** of the expression, and the constants which multiply each of these terms are called **coefficients**. So in the example above, the highest power is 3 so this expression is a polynomial of order 3. We can also note that

- the coefficient of x^3 is 5
- the coefficient of x^2 is 2
- the coefficient of x is 3

The final value of 10 is called the **constant term**. Note that any of these terms can be missing in a particular example, so the expression $2x^4 + 3x$ is also an example of a polynomial expression, although it does not contain terms in x^3 or x^2, nor a constant.

Special names are given to polynomials with the lowest orders. These are:

- **linear** for an expression of order 1
- **quadratic** for an expression of order 2, and
- **cubic** for an expression of order 3.

Examples of graphical representations of each of these are shown in Figure 3.2.

Worked Example 3.5

The heat capacities C_p for the species involved in the reaction

$$H_{2(g)} + \tfrac{1}{2}O_{2(g)} \rightarrow H_2O_{(g)}$$

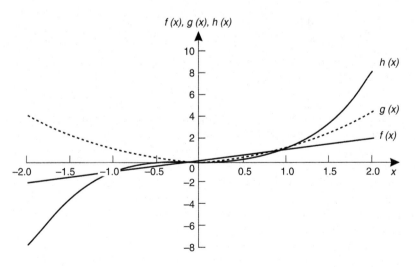

FIGURE 3.2   Graphs of representative linear $f(x)$, quadratic $g(x)$ and cubic $h(x)$ functions.

can be expressed as a polynomial with the variable T representing absolute temperature as

$$C_p = a + bT + cT^2$$

where a, b and c are constants. Use the values in the table below to obtain an expression for the change in heat capacity in terms of temperature when this reaction takes place.

	a $JK^{-1}mol^{-1}$	b $10^3\ JK^{-2}mol^{-1}$	c $10^7\ JK^{-3}mol^{-1}$
$H_{2(g)}$	29.07	−0.836	20.11
$O_{2(g)}$	25.72	12.98	−38.6
$H_2O_{(g)}$	30.36	9.61	11.8

CHEMICAL BACKGROUND

As well as the form of polynomial shown here, an expression such as

$$C_p = a + bT + \frac{c}{T^2}$$

can be used. This expression is not a polynomial because of the final term, but it can be handled in exactly the same way as the expression in this problem.

Such expressions are able to fit experimental data to within about 0.5% over a wide range of temperatures.

A mixture of hydrogen and oxygen gases will only react to give steam (gaseous water) if a spark is applied. This has the effect of supplying the activation energy for the reaction, which must always be considered in addition to arguments based on enthalpy changes.

Solution to Worked Example 3.5

This is the first problem we have met in which we have been required to extract data from a table. The form of the headings used here may seem rather strange, but this is the best way of presenting data. Let us look at the value of b for $O_{2(g)}$, the value in the very centre of the table. All we have to do is to equate the pure number in the table with its column heading, to give

$$\frac{b}{10^3\ JK^{-2}\ mol^{-1}} = 12.98$$

If we multiply each side of this equation by the units, we obtain

$$b = 12.98 \times 10^3\ J\ K^{-2}mol^{-1}$$

and, if required, we can convert to correct scientific notation by rewriting 12.98 as 1.298×10, i.e. a number between 1 and 10 multiplied by the appropriate power of 10, so that

$$b = 1.298 \times 10 \times 10^3 \, \text{J} \, \text{K}^{-2} \text{mol}^{-1}$$

$$= 1.298 \times 10^4 \, \text{J} \, \text{K}^{-2} \text{mol}^{-1}$$

A similar technique is used to read values from graphs, whose axes labels should be given similarly to these table headings.

The answer to this problem will consist of another polynomial expression whose terms represent the difference in heat capacities between the reactants and the products. It will be of the form

$$\Delta C_p = \Delta a + T \Delta b + T^2 \Delta c$$

where

$$\Delta C_p = \Sigma \, C_p \, (\text{products}) - \Sigma \, C_p \, (\text{reactants})$$

Notice that we are using notation (Δ and Σ) here to make the working more concise. Remember that Δ is used to represent a change, and that this is always taken as the final value minus the initial value. In this case, that is equivalent to taking products (final state) minus reactants (initial state). You may recall from Section 2.4.3 on page 44 that the Σ symbol means 'add up', so we are summing the contributions from the products and from the reactants for each of the coefficients a, b and c. Therefore, we have

$$\Delta a = \Sigma \, a \, (\text{products}) - \Sigma \, a \, (\text{reactants})$$

$$= 30.36 - [29.07 + (0.5 \times 25.72)] \, \text{J} \, \text{K}^{-1} \text{mol}^{-1}$$

$$= -11.57 \, \text{J} \, \text{K}^{-1} \text{mol}^{-1}$$

Note that the contribution from $O_{2(g)}$ is multiplied by 0.5 since this species appears in the reaction equation with a stoichiometry coefficient of ½. Also

$$\Delta b = \Sigma b \, (\text{products}) - \Sigma b \, (\text{reactants})$$

$$= 9.61 - [-0.836 + (0.5 \times 12.98)] \times 10^3 \, \text{J} \, \text{K}^{-2} \text{mol}^{-1}$$

$$= 3.96 \times 10^3 \, \text{J} \, \text{K}^{-2} \text{mol}^{-1}$$

and

$$\Delta c = \Sigma c \, (\text{products}) - \Sigma c \, (\text{reactants})$$

$$= 11.8 - [20.1 + (0.5 \times -38.6)] \times 10^7 \, \text{J} \, \text{K}^{-3} \text{mol}^{-1}$$

$$= 11.0 \times 10^7 \, \text{J} \, \text{K}^{-3} \text{mol}^{-1}$$

It is slightly neater to write this using correct scientific notation as

$$\Delta c = 1.10 \times 10^8 \text{J K}^{-3}\text{mol}^{-1}$$

Now substituting into our equation for ΔC_p gives

$$\Delta C_p = -11.57 \text{ J K}^{-1}\text{mol}^{-1} + (3.96 \times 10^3 \text{ J K}^{-2}\text{mol}^{-1})T$$

$$+ (1.10 \times 10^8 \text{ J K}^{-3}\text{mol}^{-1})T^2$$

Note the use of brackets to simplify this expression when the units are included. Another way of doing this would be to divide through by the common unit of J K^{-1}mol^{-1} and write

$$\frac{\Delta C_p}{\text{J K}^{-1}\text{mol}^{-1}} = -11.57 + (3.96 \times 10^3 \text{ K}^{-1})T + (1.10 \times 10^8 \text{ K}^{-2})T^2$$

3.4.3 FUNCTIONS

So far in this section, we have seen that the heat capacity can be regarded as a constant in some circumstances, while in others we can specify a dependence on the temperature. In the latter case, we would say that the heat capacity is a function of temperature. It can then be denoted as $C_p(T)$, where the T in brackets indicates that the value of T determines that of C_p. In general, we use the notation $f(x)$ to denote a function f of the variable x.

As an example, if $f(x)$ is defined as

$$f(x) = 8x^2 + 2x + 9$$

we can calculate the value of the function for any value of x. Using this notation, $f(3)$ would be the value of this function when $x=3$, and would be calculated by substituting for x on both sides of the defining equation to give

$$f(3) = (8 \times 3^2) + (2 \times 3) + 9$$

$$= (8 \times 9) + (2 \times 3) + 9$$

$$= 72 + 6 + 9$$

$$= 87$$

Worked Example 3.6

If the heat capacity C_p of solid lead is given by the expression

$$C_p(T) = 22.13 \text{ J K}^{-1}\text{mol}^{-1} + (1.172 \times 10^{-2} \text{ J K}^{-2}\text{mol}^{-1})T$$

$$+ \frac{(9.6 \times 10^4 \text{ J K mol}^{-1})}{T^2}$$

determine the heat capacity at 298 K.

CHEMICAL BACKGROUND

Observation shows that different quantities of heat are required to raise the temperature of equal amounts of different substances to the same extent. More careful experiments show that this heat capacity actually varies with the temperature. The SI unit of molar heat capacity is J K^{-1}mol^{-1}; heat changes can be measured accurately by electrical means and 1 J=1 V×l C, i.e. one joule is the work required to move an electric charge of 1 C through a potential difference of 1 V.

Solution to Worked Example 3.6

This question requires us to calculate C_p (298 K), which is given by a straightforward substitution into both sides of the defining equation.

$$C_p(298\ \text{K}) = 22.13\ \text{J K}^{-1}\text{mol}^{-1} + (1.172 \times 10^{-2}\ \text{J K}^{-2}\text{mol}^{-1} \times 298\ \text{K})$$

$$+ \frac{(9.6 \times 10^4\ \text{J K mol}^{-1})}{(298\text{K})^2}$$

$$= 22.13\ \text{J K}^{-1}\text{mol}^{-1} + 3.492\ \text{J K}^{-1}\text{mol}^{-1} + 1.081\ \text{J K}^{-1}\text{mol}^{-1}$$

$$= (22.13 + 3.492 + 1.081)\ \text{J K}^{-1}\text{mol}^{-1}$$

$$= 26.70\ \text{J K}^{-1}\text{mol}^{-1}$$

3.5 CLAPEYRON EQUATION

There is a certain amount of vapour above every solid and liquid and if this is not free to escape, as in a closed vessel, equilibrium will be reached so that as many molecules leave the surface as re-enter. At this equilibrium, the pressure above the solid or liquid is known as the vapour pressure. A study of vapour pressure is important because it can provide us with information about the solid or liquid.

The Clapeyron equation applies to any phase transition of a pure substance and is expressed by the equation

$$\frac{dp}{dT} = \frac{\Delta_t H}{T\Delta_t V}$$

where $\Delta_t H$ and $\Delta_t V$ are the enthalpy and volume changes respectively which accompany a phase transition t at absolute temperature T. The left-hand side introduces a notation that we have not met before; it needs to be treated as a complete quantity and the letter d, on its own, has no meaning in this equation. The quantity

$$\frac{dp}{dT}$$

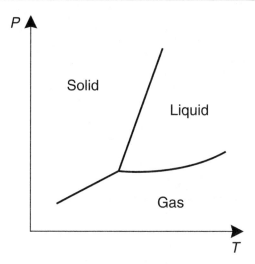

FIGURE 3.3   A typical phase diagram.

is called the **derivative of p with respect to T**. As we will see below, this function is equal to the gradient of the tangent of the graph of p against T.

The graph of pressure against temperature for a substance is known as its phase diagram and can be used to indicate where a solid, liquid or gas will be found for given values of these variables. A typical one-component phase diagram is shown in Figure 3.3.

3.5.1 DIFFERENTIATION

If we consider a function such as

$$f(x)=3x^2+1$$

whose graph is shown in Figure 2.4, we can calculate an average rate of change between any two values of x, such as $x=1$ and $x=5$. This is found by dividing the difference between the values of the function at the two values of x by the difference in the two values of x:

$$\text{average rate of change} = \frac{f(5)-f(1)}{5-1}$$

since

$$f(1)=(3\times 1^2)+1 = 3+1 = 4$$

and

$$f(5)=(3\times 5^2)+1 = 75+1 = 76$$

we have

$$\text{average rate of change} = \frac{76-4}{5-1}$$

$$= \frac{72}{4}$$

$$= 18$$

This is equivalent to calculating the gradient of the line drawn between $f(1)$ and $f(5)$ on the graph shown in Figure 3.4.

It is also possible to determine the instantaneous rate of change of this function at a specified value of x, say $x=2$. This is equivalent to determining the gradient of the tangent drawn to the curve at this value, as shown in Figure 3.5.

More generally, suppose that we wish to find the instantaneous rate of change, or gradient, when x has the specified value x_0. This is specified as

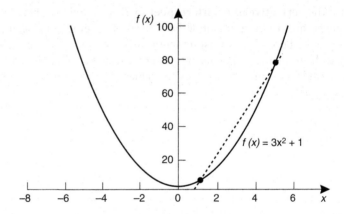

FIGURE 3.4   The graph of $f(x)=3x^2+1$.

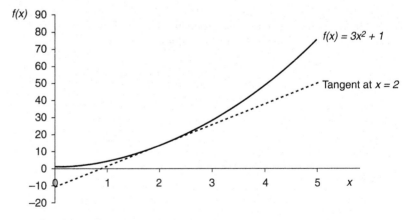

FIGURE 3.5   The gradient of the tangent to the function $f(x)=3x^2+1$, at the point where $x=2$.

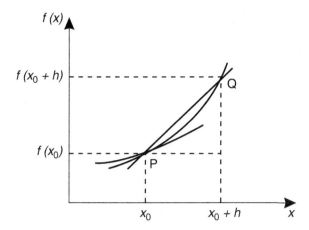

FIGURE 3.6   Definition of the derivative $\dfrac{df(x)}{dx}$.

point P in Figure 3.6. A second point Q on the curve defined by $f(x)$ has $x=x_o+h$. The gradient of the line drawn between P and Q will therefore be

$$\frac{f(x_o+h)-f(x_o)}{(x_o+h)-x_o}=\frac{f(x_o+h)-f(x_o)}{h}$$

Imagine now that Q slides down the curve until it is almost at the same place as P This has the effect of making the value of h very small. In our minds, we can make h as small as we like as long as it does not become exactly zero. This is because if h were to become zero, the denominator would be zero, but division by zero is not defined. However, making h as close to zero as possible will give a very close approximation to the value of the gradient which we require. In mathematical terminology, we would say that we wish to calculate the above expression in the limit as h tends to zero. This is written as

$$\text{Gradient of tangent} = \underset{h\to 0}{\text{Lim}}\ \frac{f(x_o+h)-f(x_o)}{h} \text{ when } x = x_o.$$

More usually, we would write this as

$$\frac{df(x)}{dx}=\underset{h\to 0}{\text{Lim}}\ \frac{f(x_o+h)-f(x_o)}{h}$$

where

$$\frac{df(x)}{dx}$$

is the derivative of $f(x)$ with respect to x, in this case when $x=x_o$. This is spoken as 'df of x by $d\,x$'. It is important to remember that this represents the gradient of a graph of $f(x)$ against x at $x=x_o$.

So far, the method of calculating a derivative involves taking a limit of a mathematical expression. In practice, this is not necessary as we can use a number of standard derivatives. In the case of a power of x, the derivative of x^n is nx^{n-1}, where n is a constant.

The following table shows this rule for $n=-2, -1, 0, 1, 2, 3, 4$.

n	x^n	Derivative
-2	x^{-2}	$-2x^{-3}$
-1	x^{-1}	$-x^{-2}$
0	$x^0=1$	$0 \times x^{-1}=0$
1	x^1	$1 \times x^0=1$
2	x^2	$2 \times x^1=2x$
3	x^3	$3 \times x^2=3x^2$
4	x^4	$4 \times x^3=4x^3$

When a power of x is multiplied by a coefficient, e.g. ax^n, the derivative is anx^{n-1} so that, for example

$$\frac{d(3x^4)}{dx}=3(4x^{4-1})=12x^3$$

Notice that a constant on its own can be thought of as a term where the constant is multiplied by an x^0, because $x^0=1$. The derivative of x^0 will always be zero, so that, for instance

$$\frac{d(6)}{dx}=\frac{d(6x^0)}{dx}=6\frac{dx^0}{dx}=6\times 0=0$$

The rule given is readily extended to more complicated functions. If, for example, we have a function $f(x)$ defined as

$$f(x)=2x^3+4x^2+7x+2$$

we can apply the standard derivatives to each of these in turn. Notice the notation used for describing each of these derivatives:

$$\frac{d(2x^3)}{dx}=2\frac{d(x^3)}{dx}=2\times 3x^2=6x^2$$

$$\frac{d(4x^2)}{dx}=4\frac{d(x^2)}{dx}=4\times 2x=8x$$

$$\frac{d(7x)}{dx}=7\frac{d(x)}{dx}=7\times 1=7$$

$$\frac{d(2)}{dx}=0$$

Putting this together gives the derivative as

$$\frac{df(x)}{dx}=6x^2+8x+7$$

Worked Example 3.7

The triple point of iodine, I_2, occurs at a temperature of 114°C and a pressure of 12 kPa. If the enthalpies of fusion $\Delta_{fus}H$ and vaporization $\Delta_{vap}H$ are 15.52 and 41.80 kJ mol^{-1} respectively, sketch the pressure-temperature graph in the region of the triple point. Take the densities of the solid and liquid to be 4.930 and 2.153 g cm^{-3} respectively and the molar mass of iodine atoms to be 126.9 g mol^{-1}. Assume that iodine vapour behaves as an ideal gas and that the enthalpy of sublimation $\Delta_{sub}H$ is given by the equation:

$$\Delta_{sub}H = \Delta_{fus}H + \Delta_{vap}H$$

CHEMICAL BACKGROUND

The term sublimation refers to a change directly from solid to vapour without passing through the liquid phase. Such behaviour can be predicted from the graph of pressure against temperature showing the equilibria between phases. From the equation given above, we can see that the enthalpy change will be the same when we pass from the solid to vapour phases, regardless of whether an intermediate liquid phase is involved.

The phase diagram obtained in this question has three lines representing phase boundaries all with positive gradients. If we repeated the exercise for water, we would find that the slope of the line representing the solid to liquid transition would be negative. This is quite unusual and is a result of the reduction in the volume of ice on melting.

Solid carbon dioxide (dry ice), naphthalene and arsenic also sublime.

Solution to Worked Example 3.7

The question gives us all the information we need to evaluate the right-hand side of the Clapeyron equation

$$\frac{dp}{dT} = \frac{\Delta_t H}{T \Delta_t V}$$

and we need to realize that the left-hand side represents the gradient of the graph of p against T. Since we are only asked about values close to the triple point, we can assume that the gradients we obtain are constant in this region of interest. We need to consider the three possible phase transitions in turn: solid to liquid; liquid to vapour; and solid to vapour.

Part 1. Solid to liquid.
The volume change ΔV will be

$$\Delta V = V_{liquid} - V_{solid}$$

where each volume can be obtained from the formula

$$V = \frac{m}{\rho}$$

with m representing mass and ρ density. Since 1 mol of I_2 has a mass of 2×126.9 or $253.8\,g$, we have

$$V_{\text{liquid}} = \frac{253.8\,\text{g mol}^{-1}}{2.153\,\text{g cm}^{-3}} = 117.9\,\text{cm}^3\,\text{mol}^{-1}$$

$$V_{\text{solid}} = \frac{253.8\,\text{g mol}^{-1}}{4.930\,\text{g cm}^{-3}} = 51.48\,\text{cm}^3\,\text{mol}^{-1}$$

so that

$$\Delta V = (117.9 - 51.48)\ \text{cm}^3\text{mol}^{-1} = 66.4\,\text{cm}^3\text{mol}^{-1}$$

We also need to convert the given temperature value of 114°C to units of K.

$$\frac{T}{K} = \frac{114°C}{°C} + 273 = 387$$

so

$$T = 387\ \text{K}$$

The Kelvin temperature scale does not use negative numbers as zero is the lowest possible temperature which can be attained. Temperatures measured on this scale are known as absolute temperatures (symbol T) and are related to those on the Celsius scale (symbol θ) by the expression $\dfrac{T}{K} = \dfrac{\theta}{°C} + 273.15$.

Note that the more precise conversion factor (273.15 rather than 273) is not used because the temperature is only given to the nearest °C.

Substituting into the Clapeyron equation now gives

$$\frac{dp}{dT} = \frac{15.52\,\text{kJ mol}^{-1}}{387\,\text{K} \times 66.4\,\text{cm}^3\,\text{mol}^{-1}}$$

$$= \frac{15.52 \times 10^3\ \text{J mol}^{-1}}{387\ K \times 66.4 \times 10^{-6}\ \text{m}^3\,\text{mol}^{-1}}$$

$$= 6.04 \times 10^5\ \text{Pa K}^{-1}$$

since $1\,\text{kJ} = 10^3\,\text{J}$, $1\,\text{cm}^3 = 10^{-6}\text{m}^3$, and $1\,\text{Pa} = 1\,\text{N m}^{-2} = 1\,\text{N m m}^{-3} = 1\,\text{J m}^{-3}$.

Note the conversion from centimetres to metres:

$$100\ \text{cm} = 1\,\text{m}$$

$$1 \text{ cm} = 10^{-2} \text{m}$$

cubing each side gives us

$$(1 \text{ cm})^3 = (10^{-2} \text{m})^3$$

or

$$1^3 \text{cm}^3 = 10^{-6} \text{m}^3$$

Notice that $(10^{-2})^3 = 10^{-6}$. Since $1^3 = 1$,

$$1 \text{cm}^3 = 10^{-6} \text{m}^3$$

Part 2. Liquid to vapour.

Since the question tells us to treat the iodine vapour as an ideal gas, we can use the ideal gas equation which will be introduced in Section 3.7 to calculate its volume at the triple point. If

$$pv = nRT$$

this can be rearranged (by dividing both sides by p) to give

$$V = \frac{nRT}{p}$$

Substituting values for 1 mol at the triple point gives us

$$V = \frac{1 \text{ mol} \times 8.314 \text{ J K}^{-1} \text{ mol}^{-1} \times 387 \text{ K}}{12 \text{ kPa}}$$

$$= \frac{8.314 \times 387}{12} \frac{\text{J}}{\text{kPa}}$$

$$= 268.126\,5 \times \frac{\text{J}}{10^3 \text{J} \text{m}^{-3}}$$

$$= 0.268\,1 \text{m}^3$$

taking the value of the gas constant R as 8.314 J K^{-1}mol^{-1} and since 1 kPa$= 10^3$ Pa$= 10^3$ J m^{-3}. We are now able to calculate the volume change for the vaporization process for 1 mol:

$$\Delta V = V_{vapour} - V_{liquid}$$

$$= 0.268\,1 \text{ m}^3 \text{mol}^{-1} - 117.9 \text{ cm}^3 \text{mol}^{-1}$$

We need to convert these two values to the same units. Generally, it is preferable to work in base units (such as metres) rather than multiples (such as centimetres) so we choose to convert cm³ to m³, and use the conversion factor $1\,\text{cm}^3 = 10^{-6}\,\text{m}^3$.

Then

$$117.9\,\text{cm}^3 = 117.9 \times 10^{-6}\,\text{m}^3$$

$$= 1.179 \times 10^{-4}\,\text{m}^3$$

If we compare this with the vapour volume of $0.268\,1\,\text{m}^3$, we see that it is negligible and we are quite justified in approximating the volume change on vaporization to be equal to the volume of the vapour. In this case, we will take ΔV as $0.268\,1\,\text{m}^3\,\text{mol}^{-1}$. Substituting into the Clapeyron equation leads us to

$$\frac{dp}{dT} = \frac{41.80\,\text{kJ}\,\text{mol}^{-1}}{387\,\text{K} \times 0.268\,1\,\text{m}^3\,\text{mol}^{-1}}$$

$$= \frac{41.80 \times 10^3\,\text{J}\,\text{mol}^{-1}}{387 \times 0.268\,1\,\text{K}\,\text{m}^3\,\text{mol}^{-1}}$$

$$= 402.9\,\text{J}\,\text{K}^{-1}\,\text{m}^{-3}$$

$$= 402.9\,\text{J}\,\text{m}^{-3}\,\text{K}^{-1}$$

$$= 402.9\,\text{N}\,\text{m}\,\text{m}^{-3}\,\text{K}^{-1}$$

$$= 402.9\,\text{N}\,\text{m}^{-2}\,\text{K}^{-1}$$

$$= 402.9\,\text{Pa}\,\text{K}^{-1}$$

Part 3. Solid to vapour.

Since the volume of the solid will be even less than that of the liquid, we are again justified in taking the volume change as being equal to the volume of the vapour, so we will set ΔV to $0.268\,1\,\text{m}^3\,\text{mol}^{-1}$. The enthalpy change $\Delta_{sub}H$ is given by the expression in the question, which evaluates to give

$$\Delta_{sub}H = \Delta_{fus}H + \Delta_{vap}H$$

$$= 15.52\,\text{kJ}\,\text{mol}^{-1} + 41.80\,\text{kJ}\,\text{mol}^{-1}$$

$$= 57.32\,\text{kJ}\,\text{mol}^{-1}$$

Substitution into the Clapeyron equation now gives

$$\frac{dp}{dT} = \frac{57.32\,\text{kJ}\,\text{mol}^{-1}}{387\,\text{K} \times 0.268\,1\,\text{m}^3\,\text{mol}^{-1}}$$

$$= \frac{57.32 \times 10^3\,\text{J}\,\text{mol}^{-1}}{387 \times 0.268\,1\,\text{m}^3\,\text{mol}^{-1}}$$

$$= 552.5\,\text{Pa}\,\text{K}^{-1}$$

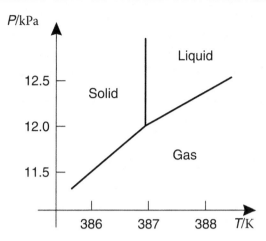

FIGURE 3.7   Phase diagram for iodine.

To summarize, the values of $\dfrac{dp}{dT}$ for the three-phase transitions are:

Part 1, solid to liquid, 6.04×105 Pa K^{-1}
Part 2, liquid to vapour, 402.9 Pa K^{-1}
Part 3, solid to vapour, 552.5 Pa K^{-1}

Using these three gradient values, we now need to draw lines of appropriate slope which pass through the triple point. Notice that the value of $\dfrac{dp}{dT}$ for the fusion process is much larger than the other two, so this can be represented by a vertical line. The value for vaporization is

$$\frac{402.9}{552.5} = 0.729$$

or 0.7 times the value for sublimation, which allows a simple qualitative representation to be drawn.

One possible graph, with values included, is shown in Figure 3.7.

3.6 CLAUSIUS-CLAPEYRON EQUATION

The Clausius-Clapeyron equation relates the vapour pressures p_1 and p_2 at two temperatures T_1 and T_2 to the enthalpy of vaporization $\Delta_{vap}H$:

$$\ln\left(\frac{p_2}{p_1}\right) = \frac{\Delta_{vap}H}{R}\left[\left(\frac{1}{T_1}\right) - \left(\frac{1}{T_2}\right)\right]$$

Here we meet the ln symbol for the first time. This is an abbreviation for natural logarithm, and this equation is telling us to 'take the natural logarithm of the quantity $\dfrac{p_2}{p_1}$.

3.6.1 LOGARITHMS

Historically, logarithms were used to make the processes of multiplication and division easier, but with the advent of electronic calculators, this use is now virtually redundant. However, their use is of much more fundamental importance than that, particularly when dealing with quantities which may span a very large range of values. There are two types of logarithms with which we need to be concerned: natural logarithms (ln), and logarithms to base 10 (log).

If three numbers a, b and c are related such that

$$a = b^c$$

then we can write

$$\log_b a = c$$

where b is called the base of the logarithm. Thus, if we choose b to have the value 10, then if

$$a = 10^c$$

it follows that

$$\log a = c$$

since we normally use the term 'log' to denote logarithms to the base 10. On the other hand, if b is set to the number known as e, which has the value 2.718 28..., we have

$$a = e^c$$

and

$$\ln a = c$$

since logarithms to the base 'e' are called natural logarithms' and are denoted by the term 'ln'.

Natural logarithms frequently arise from the mathematical analysis of problems in physical chemistry, while logarithms to the base 10 are of more use when we want to display a large range of data. The two types of logarithms are related by the expression

$$\ln x = 2303 \log x$$

We will meet an example of the natural logarithm emerging from the analysis of a physical chemistry problem in Chapter 5. This frequently arises from expressions involving a term in $\dfrac{1}{x}$, where x is a variable.

Worked Example 3.8

The vapour pressure p of neon as a function of temperature θ is as follows:

θ	p
°C	mm Hg
−228.7	19 800
−233.6	10 040
−240.2	3170
−243.7	1435
−245.7	816
−247.3	486
−248.5	325

By using an appropriate mathematical transformation, display this data graphically.

CHEMICAL BACKGROUND

These data are a selection of those which were obtained when this study was performed. The complete set was fitted to an expression of the form

$$\log\left(\frac{p}{\text{mm Hg}}\right) = \frac{0.052\,23\,A}{T} + B + CT$$

to give the values of the constants

$$A = -1\,615.5 \text{ K}$$

$$B = 5.699\,91$$

$$C = 0.011\,180\,0 \text{ K}^{-1}$$

Notice the use of the heading $\log(p/\text{mm Hg})$ in the equation above. This is because it is only possible to toke the logarithm of a pure number, i.e. one which does not have any units. If all the pressure values are divided by the units, we end up with pure numbers. For example, if $p = 19\,800$ mm Hg, we can write

$$\frac{p}{\text{mm Hg}} = \frac{19\,800\,\text{mm Hg}}{\text{mm Hg}} = 19\,800$$

which is a similar technique to the one we used in Worked Example 3.5.

Solution to Worked Example 3.8

We will consider the details of graph drawing shortly, but for the moment merely note that pressure is plotted on the vertical y-axis

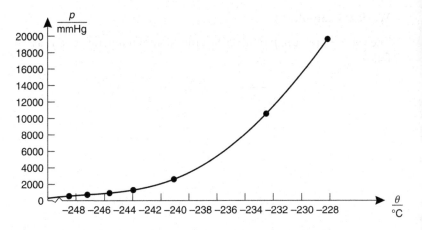

FIGURE 3.8   Graph of pressure against temperature for neon.

and temperature on the horizontal x-axis. This gives the graph shown in Figure 3.8, where we see that as the temperature increases the points become increasingly further apart and so the plot has very uneven spaces between the points. This leads to significant differences in the extent of precision to which the position of each point can be defined. In particular, it makes it very difficult to make reasonable estimates of behaviour at small values of temperature and pressure. Suppose that instead of plotting pressure, we take its logarithm to base 10 and plot that against temperature. Using a calculator gives us the following values:

$\dfrac{\theta}{°C}$	$\log\left(\dfrac{p}{mm\,Hg}\right)$
−228.7	4.297
−233.6	4.002
−240.2	3.501
−243.7	3.157
−245.7	2.912
−247.3	2.687
−248.5	2.512

Notice that the

$$\log\left(\frac{p}{mm\,Hg}\right)$$

values now occupy a much smaller range. From Figure 3.9, we can also see that they are far more evenly spread. This means that it is much easier to read off values at lower temperatures and pressures, and consequently, we can predict the behaviour of neon more reliably.

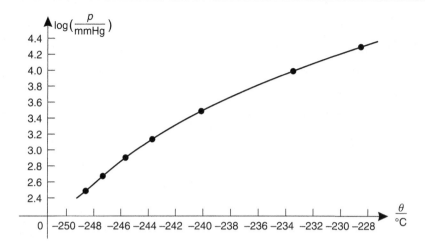

FIGURE 3.9   Graph of the logarithm of pressure against temperature for neon.

3.6.2 THE EQUATION OF A STRAIGHT LINE

We saw in Section 3.4.2 on page 65 that a polynomial of order 1 was known as a linear expression. Some examples are

$$5x+2$$

$$3x-7$$

$$0.4x-0.3$$

These can be summarized by the general equation

$$y=mx+c$$

where y is the value we evaluate on the right-hand side of the equals sign, and m and c are constants. So if we write the first example as

$$y=5x+2$$

the constant m will be 5, and c will be 2. If we were to plot a graph of y against x, we would find that it was a straight line with gradient 5 and intercept 2, as shown in Figure 3.10. The consequence of this is that if we can rearrange any mathematical relationship into the form of a linear equation, we can obtain a straight line graph. Frequently, the values of the gradient and intercept of such graphs give us useful information about the system being studied.

By convention, the term intercept is used for the point where a line crosses the y-axis. There is also (in general) an intercept where the line crosses the x-axis, but this is of less use.

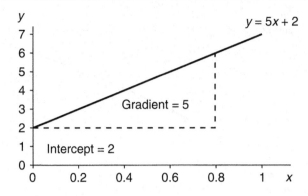

FIGURE 3.10   Graph of $y = 5x + 2$.

Worked Example 3.9

At the standard pressure, $p^{\ominus} = 1$ atm, a liquid boils at its normal boiling point, T_b. Given a set of values of the vapour pressure p at absolute temperature T, show how the Clausius-Clapeyron equation can be used to obtain the enthalpy of vaporization of the liquid.

CHEMICAL BACKGROUND

A liquid boils when its vapour pressure is equal to the atmospheric or external pressure. More molecules are then leaving the surface of the liquid than are re-entering. Consequently, as the external pressure is reduced, the boiling temperature of the liquid falls.

Pressure is frequently expressed in a number of different units. Equivalent values for atmospheric pressure are 1 atm, 760 mm Hg, and 101.325 kPa.

Solution to Worked Example 3.9

Starting from the Clausius-Clapeyron equation (see Section 3.6)

$$\ln\left(\frac{p_2}{p_1}\right) = \frac{\Delta_{vap}H}{R}\left[\left(\frac{1}{T_1}\right) - \left(\frac{1}{T_2}\right)\right]$$

we can substitute the pair of values given into the equation. It is simpler to set $p_1 = p^{\ominus}$ and $T_1 = T_b$ and then to replace the symbols p_2 and T_2 by p and T respectively. In effect, this sets p_1 and T_1 to specific pressure and temperature values, while p_2 and T_2 become the general pressure and temperature. Doing this gives the expression

$$\ln\left(\frac{p}{p^{\ominus}}\right) = \frac{\Delta_{vap}H}{R}\left[\left(\frac{1}{T_b}\right) - \left(\frac{1}{T}\right)\right]$$

At this stage, it is important to realize that p and T are the values of our variable pressure and temperature respectively, while $\Delta_{vap}H$, R, $p^{\ominus}$ and T_b are all constants. If we expand the right-hand side of the equation (by removing the square brackets) we have

$$\ln\left(\frac{p}{p^{\ominus}}\right) = \left(\frac{\Delta_{vap}H}{RT_b}\right) - \left(\frac{\Delta_{vap}H}{RT}\right)$$

which can be written as

$$\ln\left(\frac{p}{p^{\ominus}}\right) = -\frac{\Delta_{vap}H}{R}\left(\frac{1}{T}\right) + \left(\frac{\Delta_{vap}H}{RT_b}\right)$$

Consider what happens if we make the substitutions

$$x = \frac{1}{T} \qquad\qquad y = \ln\left(\frac{p}{p^{\ominus}}\right)$$

$$m = -\frac{\Delta_{vap}H}{R} \qquad\qquad c = \frac{\Delta_{vap}H}{RT_b}$$

You will find that this gives back the general equation

$$y = mx + c$$

If we plot $\ln\left(\frac{p}{p^{\ominus}}\right)$ on the vertical y-axis and $\frac{1}{T}$ on the horizontal x-axis, a straight line will be obtained with gradient $-\frac{\Delta_{vap}H}{R}$ and intercept $\frac{\Delta_{vap}H}{RT_b}$.

Since the value of the gas constant R is known, the gradient gives the value of $\Delta_{vap}H$. Notice that we need to plot $\ln\left(\frac{p}{p^{\ominus}}\right)$ on the y-axis. Since $p^{\ominus} = 1$ atm, if the values of p are also expressed in atm we will automatically be taking the natural logarithm of pure numbers, as required.

3.6.3 PLOTTING GRAPHS

Having seen how to determine the plot we require, it is worth spending some time considering the mechanics of graph plotting. While there are now many computer programs available for doing this, they need to be used carefully to avoid giving unusual results. It is also worth discussing manual graph plotting because the rules which govern overall appearance should also be applied to computer-generated plots.

If we have an equation relating two variables, we should be able to determine which variable to plot on the x- and y-axes. However, there will be times when we will be instructed to 'plot variable 1 against variable 2'. This is *always* equivalent to plotting 'y against x', and it is important to display these variables the right way around.

Having established this, the most important thing to remember is to fill as much of the graph paper as possible. It is best to have one plot per page, except when superimposing plots using the same axes for both, which can sometimes be useful. Try to avoid scales involving numbers such as 3 and 7 since these make it difficult to work out exactly where intermediate values should be plotted. Use a sharp pencil, label the axes and, if appropriate, include a title for the graph. There is no need to include the origin on either axis unless the spread of data lends itself to doing so.

Worked Example 3.10

The vapour pressure p of water is given below as a function of temperature θ.

Determine the enthalpy of vaporization of water.

$\dfrac{\theta}{°C}$	$\dfrac{p}{10^{-2}\,\text{atm}}$
20	2.31
30	4.19
40	7.28
50	12.2
60	19.7
70	30.8

CHEMICAL BACKGROUND

The vapour pressure increases with temperature as a result of intermolecular forces in the liquid being broken. Vapour pressure measurements are consequently able to give us information about the nature of those forces. Most liquids with high values of enthalpy of vaporization also have high normal boiling points, although this is not always the case.

Typical values of the standard enthalpy of vaporization are $8.2\,\text{kJ}$ mol^{-1} for methane and $43.5\,\text{kJ mol}^{-1}$ for ethanol; the latter is higher due to the presence of hydrogen bonding in the liquid.

Solution to Worked Example 3.10

We saw in the previous problem that we need to plot $\ln\left(\dfrac{p}{p^{\ominus}}\right)$

against $\dfrac{1}{T}$ where $p^{\ominus}=1$ atm. It is important to realize that T refers to the absolute temperature (in units of K) rather than the values given here. The first stage is to draw up a table containing the transformed data. If we consider the first pair of values, we have

$$\theta = 20°C$$

so that

$$T = (20 + 273) = 293 \text{ K}$$

and

$$\frac{1}{T} = \frac{1}{293\,K}$$

$$= 3.41 \times 10^{-3}\ K^{-1}$$

$$p = 2.31 \times 10^{-2}\ atm$$

so that

$$\frac{p}{p^{\ominus}} = \frac{2.31 \times 10^{-2}\,atm}{1\,atm}$$

$$= 2.31 \times 10^{-2}$$

and

$$\ln\left(\frac{p}{p^{\ominus}}\right) = \ln(2.31 \times 10^{-2})$$

$$= -3.77$$

Applying a similar treatment to the other values leads to the final table:

$\dfrac{T^{-1}}{10^{-3}\,K^{-1}}$	$\ln\left(\dfrac{p}{p^{\ominus}}\right)$
3.41	−3.77
3.30	−3.17
3.19	−2.62
3.10	−2.10
3.00	−1.63
2.92	−1.18

A suitable range of values on the x-axis (T^{-1}) would be from 2.9 to 3.5, with divisions of 0.1. On the y-axis $\left(\ln\left(\dfrac{p}{p^{\ominus}}\right)\right)$ it would be from −3.5 to −1.0, with divisions of 0.5.

Exactly how these are fitted onto the graph paper depends on the particular arrangement of ruled squares. The labels on the axes are exactly as in the table above, to ensure that the points on the graph actually represent pure numbers. To obtain the gradient, we need to measure the increase in y and divide it by the corresponding increase in x, making sure that we use as large a portion of the straight line as possible. Using the values shown on the graph in Figure 3.11, this gives us

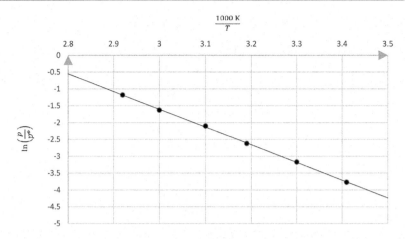

FIGURE 3.11   Graph of $\ln\left(\dfrac{p}{p^{\ominus}}\right)$ against 1000 K/T $\dfrac{1000\,\mathrm{K}}{T}$ in Worked Example 3.10.

$$\text{Gradient} = \frac{-0.6-(-4.2)}{(2.8-3.5)\times 10^{-3}\,\mathrm{K}^{-1}}$$

$$= \frac{3.6}{-0.7\times 10^{-3}\,\mathrm{K}^{-1}}$$

$$= -5.14\times 10^{3}\ \mathrm{K}$$

Notice that, in the first line of gradient calculation, the corresponding values of T^{-1} and $\ln\left(\dfrac{p}{p^{\ominus}}\right)$ appear directly under one another. We know that the value of the gradient is equal to $-\dfrac{\Delta_{\mathrm{vap}}H}{R}$ so

$$-\frac{\Delta_{\mathrm{vap}}H}{R} = -5.14\times 10^{3}\,\mathrm{K}$$

Multiplying both sides of the equation by R and multiplying both sides by −1 to change the negative signs to positive signs on both sides gives

$$\Delta_{\mathrm{vap}}H = 5.14\times 10^{3}\ \mathrm{K}\times R$$

which on substituting $R=8.314\ \mathrm{J\ K^{-1}mol^{-1}}$ gives

$$\Delta_{\mathrm{vap}}H = 5.14\times 10^{3}\ \mathrm{K}\times 8.314\ \mathrm{J\ K^{-1}mol^{-1}}$$

$$= 42.7\times 10^{3}\ \mathrm{J\ mol^{-1}}$$

This can be written more concisely by using kJ rather than 10^{3} J as

$$\Delta_{\mathrm{vap}}H = 42.7\,\mathrm{kJ\ mol^{-1}}$$

3.6.4 PLOTTING GRAPHS USING A SPREADSHEET

We will now plot the data from Worked Example 3.10 using a spreadsheet.

As in Section 2.4.3.2 on page 51, the instructions given will refer to the Microsoft Excel spreadsheet; however, those for other spreadsheets will be very similar.

Begin by putting table headings at the top of the first two columns. In cell A1, type the letter q. Right click the letter q, select Format Cells, and choose the Symbol font from the drop-down menu followed by OK. This changes the letter q to θ. In cell B1, type the letter p. We will leave a line for clarity, so the first value of θ, 20, is put into cell A3. Next to it in cell B3, type 2.31E-02; this is the spreadsheet shorthand for 2.31×10^{-2}. Similarly, we can enter the data to fill cells A4 to A8 with values of θ and cells B4– B8 with values of p. Notice that the final 3 values of p are automatically changed to standard form (Section 1.7) by the program.

The first step is to transform the data as we did manually in Section 3.6.3 on page 85. We will leave a blank column, so the transformed values of θ will be put into column D. Begin by putting the column heading T in cell D1. In D3 we type the formula=A3+273 to convert the temperature in °C to a value in K. Position the mouse at the bottom right corner of D3 and click and drag down to D8 to copy the formula. These cells are now filled by the temperatures in K. Now we move to column E and put the heading $1/T$ in cell E1. The formula in cell E3 will be=1/D3; this needs to be copied by dragging the corner of E3 down as far as E8 to fill the cells with the values of $1/T$. In cell F1 we type the column heading "ln p"; in F3 the formula=LN(B3) which is copied by dragging to all the cells up to and including F8 to fill them with the values of ln p. The spreadsheet will now look as shown in Figure 3.12.

The data to be plotted now reside in columns E and F. Highlight cell E3 and drag the mouse down to E8 and across to F8, so that the cells containing data in both columns are highlighted. From the chart tab on the Insert menu choose XY(scatter) and the icon for unconnected points.

Now click on the+symbol at the top right of the chart. Check the Trendline box which will produce a line of best fit through the points. Right click anywhere on that trendline and select "Format Trendline" from the drop-down menu. You may need to scroll down to select the option "Display Equation on Chart". This will result in the equation

$$y = -5\,208.2\,x + 14.014$$

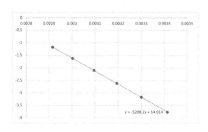

θ	p		T	1/T	ln p
20	2.31E-02		293	0.003413	-3.76792
30	4.19E-02		303	0.0033	-3.17247
40	7.28E-02		313	0.003195	-2.62004
50	1.22E-01		323	0.003096	-2.10373
60	1.97E-01		333	0.003003	-1.62455
70	3.08E-01		343	0.002915	-1.17766

FIGURE 3.12   Spreadsheet and chart for Worked Example 3.10.

being displayed. This is of the form $y=mx+c$ (see Section 3.6.2) so that the gradient m has the value −5 208.2. This spreadsheet treatment does not give us any information about the units, so we need to be aware from our knowledge of the problem that the gradient is −5 208.2 K.

The resulting spreadsheet and chart are shown in Figure 3.12.

We now proceed as in the previous section.

$$-\frac{\Delta_{vap}H}{R}=-5\,208.2\,K$$

Rearranging this gives

$$\Delta_{vap}H=5\,208.2\,K\times8.314\,J\,K^{-1}mol^{-1}$$

$$=43.3\times10^{3}J\,mol^{-1}$$

$$=43.3\,kJ\,mol^{-1}$$

3.7 THE IDEAL GAS EQUATION

The ideal gas equation is the relationship you will probably meet most frequently in thermodynamics. This is because it describes the simplest state of matter to study, and is itself a simple expression. The volume of the molecules in an ideal gas would be negligible as would the forces between them. Real gases behave like ideal gases and obey the ideal gas equation at low pressures and high temperatures.

The ideal gas equation usually appears in the form

$$pV=nRT$$

where p is the pressure, V the volume, n the amount of gas, R the ideal gas constant and T the absolute temperature.

3.7.1 DIMENSIONAL ANALYSIS

At its simplest level dimensional analysis allows us to check the validity of an equation. When deriving equations it is a powerful technique for suggesting relationships between variables and physical constants. It is important to note that it will not give any information about multiples of units so, for example, we are unable to distinguish between mm, cm, m and km. In this case, we would work with the base unit of length which is m. You will find Appendix A.2 on page 341 useful for showing the relationship between the units we will commonly meet.

Worked Example 3.11

Show that the ideal gas equation $pV=nRT$ is dimensionally correct.

CHEMICAL BACKGROUND

This is a simple but useful example as it contains a reasonable number of quantities which are multiplied. Once we have sums and differences, we have to check that we are adding quantities with the same units.

Solution to Worked Example 3.11

The first step is to identify the units of each of the quantities.

Pressure p is measured in Pa. We see from Appendix A.2 that this is equivalent to N m^{-2} or kg m^{-1}s^{-2}.

Volume is typically measured in dm^3. We can ignore the prefix and assume that the unit is m^3.

The variable n is the amount of substance measured in mol.

R is the gas constant 8.314 J K^{-2}mol^{-1}. We see from Appendix A.2 that J is equivalent to N m or kg m^{2}s^{-2}.

T is the temperature in K.

On the left-hand side of the equation, the units of pV are

$$N\,m^{-2}m^3$$

and if we combine these using the rules of indices (Section 1.6.3 on page 18) we have

$$N\,m^{(-2+3)} = N\,m$$

which as we saw above is equivalent to J.

On the right-hand side, the units of nRT will be

$$mol\,J\,K^{-1}mol^{-1}\,K$$

Using the rules of indices shows that mol mol^{-1} is mol^0 and any number raised to power zero is 1. Similarly

$$K^{-1}\,K = K^0 = 1$$

We are then left with units of J. Since the units are the same on both sides of the equation it is dimensionally correct.

Notice that we could have used different equivalences for Pa and J but since they appear on opposite sides of the equation we were able to avoid this complication.

3.7.2 INTERCONVERSION OF UNITS

We have already met several examples in which units have been substituted along with the corresponding numerical value in an equation. This practice is essential and allows us to obtain the correct units of the final calculated quantity in a systematic way. A similar technique may also be used to convert a quantity from one set of units to another.

We have already seen that for practical purposes we may wish to work with the pressure units of atm rather than kPa. Other non-SI units in common use are the angstrom (Å) which is equal to 10^{-10} m and the calorie (cal) which is equal to 4.184 J. The use of such units in certain fields is so well established that to try to change them would probably be more confusing than learning to live with them.

Worked Example 3.12

The gas constant R is usually given as 8.314 J $K^{-1}mol^{-1}$. What is its value when expressed in the units dm^3 atm $K^{-1}mol^{-1}$?

CHEMICAL BACKGROUND

The value of the gas constant was formerly determined from measurements of the molar volumes of oxygen and nitrogen, and more recently by determining the speed of sound in argon. Since 2019 it has an exact value defined as the Avogadro constant L multiplied by the Boltzmann constant k. Such changes in the way in which the values of constants are determined to do take place from time to time, but invariably they lead to very small changes in their numerical values which are only significant in high precision work.

Solution to Worked Example 3.12

To solve this problem, we need to use the following conversion factors:

$$1 \text{ atm} = 101.325 \text{ kPa} = 101.325 \times 10^3 \text{ Pa}$$

$$1 \text{ J} = 1 \text{ N m}$$

$$1 \text{ Pa} = 1 \text{ Nm}^{-2}$$

Comparing the units of the two forms of R shows that we essentially have to convert J to dm^3 atm. We do this by successive substitution in the original expression for R.

$$R = 8.314 \text{ J } K^{-1}mol^{-1}$$

$$= 8.314 \text{ N m } K^{-1}mol^{-1}$$

Notice that to obtain units of atm, we need to convert from Pa. At this stage, it helps to realize that

$$1 \text{ Pa} = 1 \text{ N m}^{-2} \text{ and } 1 \text{ J} = 1 \text{ N m}^{-2} \text{m}^3$$

so that

$$R = 8.314 \text{ N m}^{-2} \text{m}^3 \text{ } K^{-1}mol^{-1}$$

$$= 8.314 \text{ Pa m}^3 \text{ } K^{-1}mol^{-1}$$

Rearranging the definition for 1 atm and substituting 10 dm=1 m gives us

$$R = \frac{8.314\,\text{atm}}{101.325 \times 10^3} \times (10\,\text{dm})^3\,\text{K}^{-1}\,\text{mol}^{-1}$$

$$= 8.205 \times 10^{-5}\,\text{atm} \times 10^3\,\text{dm}^3\,\text{K}^{-1}\text{mol}^{-1}$$

$$= 8.205 \times 10^{-2}\,\text{dm}^3\,\text{atm}\,\text{K}^{-1}\text{mol}^{-1}$$

3.7.3 CONSTANTS AND VARIABLES

The gas constant R is known as a **universal constant**, which means that its value does not change under any circumstances. This is in contrast to quantities such as n in the ideal gas equation. This is the amount of gas, which will be fixed for a given sample of gas in a closed container, but which may vary from sample to sample. The quantities p, V and T are variables, but any of them may be considered to be fixed according to experimental conditions.

Worked Example 3.13

Identify the constants and variables in the following equations which also describe the behaviour of gases.

a. The van der Waals equation is

$$\left(p + \frac{an^2}{V^2}\right)(V - nb) = nRT$$

Here p is the pressure of gas, V its volume, T the temperature and n the amount present. For carbon monoxide, $a = 0.150\,5$ Pa m^{6}mol^{-2} and $b = 0.039\,8 \times 10^{-3}$m^{3}mol^{-1}.

b. The Beattie-Bridgeman equation is

$$p = \frac{RT\left[1 - \left(\dfrac{c}{V_m T^3}\right)\right]}{V_m^2}(V_m + B) - \frac{A}{V_m^2}$$

with

$$A = A_o\left[1 - \left(\frac{a}{V_m}\right)\right]$$

and

$$B = B_o\left[1 - \left(\frac{b}{V_m}\right)\right]$$

Here p is the pressure of gas, V_m its molar volume and T the temperature. For methane, $A_o = 0.230\,71\ \text{Pa m}^6\text{mol}^{-1}$, $a = 1.855 \times 10^{-5}\ \text{m}^3\ \text{mol}^{-1}$, $B_o = 5.587 \times 10^{-5}\ \text{m}^3\ \text{mol}^{-1}$, $b = -1.587 \times 10^{-5}\text{m}^3\ \text{mol}^{-1}$, $c = 128.3\ \text{m}^3\ \text{mol}^{-1}$.

c. The virial equation is

$$pV_m = RT + Bp$$

Here p is the pressure of gas, V_m its molar volume, R the gas constant and T the absolute temperature. For nitrogen, the second virial coefficient B is $-4.2\,\text{cm}^3\text{mol}^{-1}$ at 300 K.

Note that the constant B in (b) and (c) is different.

CHEMICAL BACKGROUND

a. The van der Waals equation was developed to account for the behaviour of a gas when it is liquefied. The volume term is reduced to allow for the finite volume of the molecules, while the pressure term is increased to allow for the presence of forces between molecules.

b. The inclusion of a large number of parameters in the Beattie-Bridgeman equation allows its use in situations where a precise fit to experimental data is required, particularly at high pressures.

c. The form of the virial equation given here actually comes from a power series of the form

$$pV_m = RT + Bp + Cp^2 + Dp^3 + \ldots$$

where B, C and D are functions of temperature and are known as the second, third and fourth virial coefficients respectively. One of the uses of this equation is that these coefficients may be related to the potential functions which describe the interaction between molecules. It is common to ignore terms in p^2 and higher.

Solution to Worked Example 3.13

a. As in the case of the ideal gas equation, the variables are p, V and T, if the amount of gas n is fixed. The values of a and b are constant, depending only on the nature of gas being considered.

b. This is also similar to the ideal gas equation in that p, V_m and T may vary. The other quantities A_o, a, B_o, b and c are constants and again depend only on the nature of the gas.

c. As previously p, V_m and T are variables and R is a constant. However, the question suggests that B is a function of temperature so this could also be regarded as a variable, although its fixed value at a given temperature may be obtained from tables.

3.7.4 PROPORTION

We can distinguish two types of proportion. If two quantities are **directly proportional** to one another, the ratio of their values will remain constant. For example, if x and y can have the following pairs of values they are in direct proportion:

x	1	2	3	4
y	2	4	6	8

In this case, the ratio $\dfrac{y}{x}$ remains constant at 2. We say that y is directly proportional to x. This is written mathematically as

$$y \propto x$$

where $\propto$ is the proportionality sign. If this is the case, we can introduce a constant of proportionality K and write the relationship as

$$y = Kx$$

Rearranging this equation gives

$$K = \frac{y}{x}$$

and we see that the constant of proportionality is the constant ratio of the two quantities x and y.

Conversely, x and y may be **inversely proportional** to one another. An example of such a relationship is in the following table:

x	1	2	3	4
y	24	12	8	6

This time, the product xy remains constant with a value of 24. In this case, we would write

$$y \propto \frac{1}{x}$$

which on replacing a by a new proportionality constant K' gives

$$y = \frac{K'}{x}$$

which rearranges to

$$K' = xy$$

so K' is the constant product of x and y.

Worked Example 3.14

For a fixed amount of gas which obeys the ideal gas equation $pV=nRT$, describe the relationship between the two variables indicated when the third is held constant:

a. p and V at constant T
b. p and T at constant V
c. T and V at constant p.

CHEMICAL BACKGROUND

The experimental relationships between these variables are expressed by Boyle's Law and Charles's Law. Together with Avogadro's Law, these can be used to formulate the ideal gas equation. This problem involves deducing the original relationships from the final equation.

Boyles's Law states that pressure is inversely proportional to volume, Charles's Law that volume is proportional to temperature and Avogadro's Law that the volume is proportional to the amount of gas present.

Solution to Worked Example 3.14

To solve this problem we need to be able to rearrange the ideal gas equation

$$pV=nRT$$

so that the two variables are on opposite sides of the equation. In each case, we will call the grouping of constant terms K, which will be our constant of proportionality.

a. Dividing both sides of the equation by V gives

$$p = \frac{nRT}{V}$$

and since nRT is a constant, which we will call K, we can write

$$p = \frac{K}{V}$$

or

$$p \propto \frac{1}{V}$$

so p and V are inversely proportional to one another. This is Boyle's Law.

b. Using the rearranged equation from part (a)

$$p = \frac{nRT}{V}$$

and since $\dfrac{nR}{V}$ is a constant, which we will call K', we can write

$$p=K'T$$

or

$$p \propto T$$

so p and T are directly proportional.

c. Dividing both sides of the original equation by p gives

$$V = \frac{nRT}{p}$$

and since $\dfrac{nR}{p}$ is a constant, which we will call K'', we can write

$$V=K''T$$

or

$$V \propto T$$

so V and T are directly proportional. This is Charles's Law.

3.7.5 FUNCTIONS OF TWO VARIABLES

In Section 3.4.3 on page 69, we met the idea of writing an expression as a function of a variable, using the $f(x)$ notation. This can be extended to cases where we have two (or more) variables, such as

$$f(x, y)=x^3+3x^2y+2y^2$$

and

$$g(x, y)=\ln(3x)+2xy$$

Note that we can also write expressions such as $z=f(x, y)$ and $z=g(x, y)$ in order to facilitate the plotting of such functions in three dimensions.

Worked Example 3.15

Use the function notation to write expressions for the

a. pressure
b. volume, and
c. temperature of an ideal gas.

CHEMICAL BACKGROUND

Whereas we can display a function of one variable as a line graph, when two variables are involved, this is not possible. One solution is to use a three-dimensional plot, where the height above the axes gives the value of the functions. This has been done for this example in Figure 3.13.

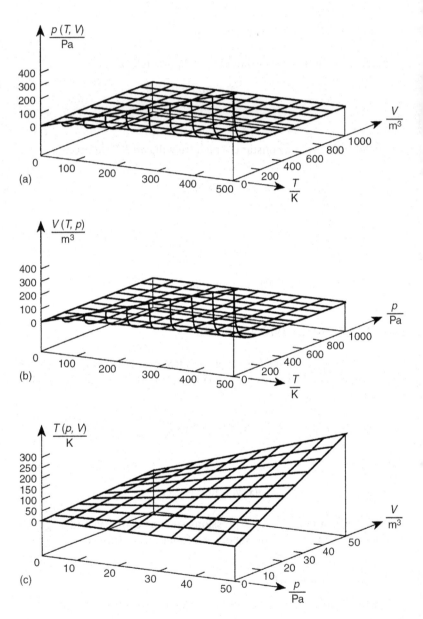

FIGURE 3.13   Plots representing the ideal gas equation. Units are Pa for pressure p, m^3 for volume V and K for absolute temperature T. (a) p as a function of T and V (b) V as a function of p and T (c) T as a function of p and V.

Solution to Worked Example 3.15

a. We can use the rearranged equation from part (a) of Worked Example 3.14. Since both temperature and volume are variables, we write

$$p(T,V) = \frac{nRT}{V}$$

b. We can use the rearranged equation from part (c) of Worked Example 3.14. Since both temperature and pressure are variables, we write

$$V(T,p) = \frac{nRT}{p}$$

c. We can rearrange the ideal gas equation (by dividing both sides by nR) to make T the subject:

$$T = \frac{pV}{nR}$$

Since p and V are both variables

$$T(p,V) = \frac{pV}{nR}$$

3.7.6 PARTIAL DIFFERENTIATION

We have already met the idea of differentiating a function of one variable in Section 3.5.1 on page 71. We saw that this was equivalent to determining the gradient of the curve representing that function, and represented the change in the function for a given change in the variable.

When we have a function of two variables, changing either of them will produce a change in the value of the function. This change is given by a quantity called the **partial derivative**. If our function is $f(x, y)$, then it is possible to determine

$$\left(\frac{\partial f(x,y)}{\partial x} \right)_y$$

the partial derivative with respect to x when y is held constant and

$$\left(\frac{\partial f(x,y)}{\partial y} \right)_x$$

the partial derivative with respect to y when x is held constant. Note that, in both cases, the quantity to be held constant is shown outside the bracket. The notation for each derivative is often abbreviated to

$$\frac{\partial f}{\partial x} \text{ and } \frac{\partial f}{\partial y}$$

respectively for clarity and to save space. We then assume that it is the variable which does not appear in the derivative that remains constant.

The quantity $\dfrac{\partial f}{\partial x}$ represents the change in $f(x, y)$ when x is changed and is equal to the gradient of the tangent to the surface of $f(x, y)$ in the x-direction, as shown in Figure 3.14 where the tangent is drawn at a constant value of y, denoted as k. Similarly, $\dfrac{\partial f}{\partial y}$ represents the change in $f(x, y)$ when y is changed and is equal to the gradient of the tangent to the surface $f(x, y)$ in the y-direction.

Partial derivatives are calculated in the same way as full derivatives while holding the fixed variable constant. If a function is defined so that

$$f(x, y) = 2x^2 + 3xy + 2xy^2$$

we can calculate partial derivatives with respect to both x and y.

In the first case, we can rewrite this treating every term in y as a constant. In the expression below, all the constant terms are enclosed in brackets:

$$f(x, y) = (2)x^2 + (3y)x + (2y^2)x$$

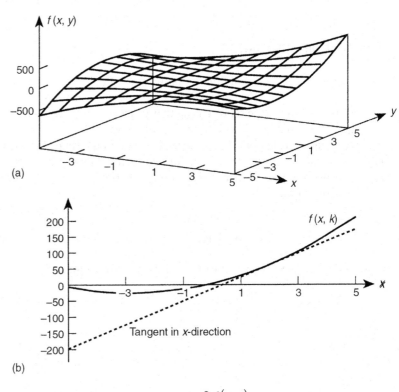

(a)

(b)

FIGURE 3.14 The partial derivative $\dfrac{\partial f(x,y)}{\partial x}$ at $y=k$. (a) f as a function of x and y (b) f as a function of x when $y=k$.

Differentiating x^2 gives $2x$ and differentiating x gives 1, so we obtain

$$\frac{\partial f}{\partial x} = 2(2x) + (3y) \times 1 + (2y^2) \times 1$$

$$= 4x + 3y + 2y^2$$

Similarly, if we rewrite the expression for $f(x, y)$ treating every term in x as a constant then

$$f(x, y) = (2x^2) + (3x)y + (2x)y^2$$

Differentiating a constant gives 0, differentiating y gives 1 and differentiating y^2 gives $2y$. We therefore obtain

$$\frac{\partial f}{\partial y} = 0 + (3x) \times 1 + (2x) \times 2y$$

$$= 3x + 4xy$$

Worked Example 3.16

For an ideal gas, calculate the partial derivatives:

$$\frac{\partial p}{\partial V} \text{ and } \frac{\partial p}{\partial T}$$

CHEMICAL BACKGROUND

Partial derivatives find much use in thermodynamics, as it is possible to have relationships between them. One example of this is Maxwell's relations, which consist of four relationships of the type

$$\left(\frac{\partial V}{\partial T} \right)_p = -\left(\frac{\partial S}{\partial p} \right)_T$$

These are useful because, while it is easy to measure the variation of volume V with temperature T, it is not straightforward to measure entropy S as a function of pressure p in the laboratory.
    The other Maxwell relations are:

$$\left(\frac{\partial T}{\partial V} \right)_S = -\left(\frac{\partial p}{\partial S} \right)_V$$

$$\left(\frac{\partial T}{\partial p} \right)_S = \left(\frac{\partial V}{\partial S} \right)_p$$

$$\left(\frac{\partial p}{\partial T} \right)_V = \left(\frac{\partial S}{\partial V} \right)_T$$

Solution to Worked Example 3.16

We can use the expression obtained in part (a) of Worked Example 3.15:

$$p(T,V) = \frac{nRT}{V}$$

To calculate the partial derivative with respect to V, we treat T as a constant, so the expression on the right-hand side can be written with the constant term in brackets as

$$p(T,V) = (nRT) \times \frac{1}{V}$$

$$= (nRT)V^{-1}$$

Remembering that we differentiate by multiplying by the power and reducing the power by 1, x^{-1} would be differentiated (with respect to x) to give $(-1)x^{-2}$, i.e.

$$\frac{d}{dx}(x^{-1}) = (-1)x^{-1-1} = -\frac{1}{x^2}$$

Therefore in this case, we obtain

$$\frac{\partial p}{\partial V} = nRT\left(-\frac{1}{V^2}\right)$$

$$= -\frac{nRT}{V^2}$$

To calculate the partial derivative with respect to T, we treat V as a constant, so grouping the constant term in brackets gives us

$$p(T,V) = \left(\frac{nR}{V}\right)T$$

Remember that differentiating x with respect to x gives us 1, i.e.

$$\frac{dx}{dx} = 1$$

Here we are differentiating with respect to T, which gives

$$\frac{\partial p}{\partial T} = \left(\frac{nR}{V}\right) \times 1$$

$$= \frac{nR}{V}$$

3.7.7 THE DIFFERENTIAL

We saw, in Worked Example 3.10, that we could determine the gradient m of a straight line graph of y against x as

$$m = \frac{\text{increase in } y}{\text{increase in } x} = \frac{\Delta y}{\Delta x}$$

It is possible to rearrange this equation to give an expression for the increase in either of the variables x or y:

$$\Delta x = \frac{\Delta y}{m}, \Delta y = m \Delta x$$

Similarly, we saw in Section 3.5.1 on page 71 that the gradient of a curve is given by the derivative $\frac{dy}{dx}$ which comes from considering very small or infinitesimal changes in the variables x and y. By analogy with our treatment for the straight line above, we can write

$$dy = m \, dx$$

where the gradient m will be a function of x. This expression allows us to express the change in variable y in terms of the variable x and the change in x.

For example, if

$$y = 2x^2 + x + 3$$

the derivative is

$$\frac{dy}{dx} = 4x + 1$$

and so

$$dy = (4x + 1) \, dx$$

where dy is known as the **differential.**

This approach can be readily extended to functions of more than one variable, such as $f(x, y)$. The differential $df(x, y)$ is then defined by

$$df(x, y) = \left(\frac{\partial f}{\partial x} \right)_y dx + \left(\frac{\partial f}{\partial y} \right)_x dy$$

This can be interpreted as giving the change in the function $f(x, y)$ when the variables x and y are changed by small amounts dx and dy, respectively. For example, if

$$f(x, y) = xy^3 + 2x^2y$$

we calculate the partial derivative $\dfrac{\partial f}{\partial x}$ by treating y as a constant. Then

$$f(x, y) = (y^3)x + (2y)x^2$$

and

$$\frac{\partial f}{\partial x} = (y^3) \times 1 + (2y) \times 2x$$

$$= y^3 + 4xy$$

Treating x as a constant allows us to calculate the partial derivative $\dfrac{\partial f}{\partial y}$:

$$f(x, y) = (x)y^3 + (2\ x^2)y$$

and

$$\frac{\partial f}{\partial y} = (x) \times 3y^2 + (2x^2) \times 1$$

$$= 3xy^2 + 2x^2$$

Substituting in our general expression above for the differential gives:

$$df = (y^3 + 4xy)\ dx + (3xy^2 + 2x^2)\ dy$$

Note that for simplicity we have replaced the term $df(x, y)$ by df.

Worked Example 3.17

Calculate the differential dp of an ideal gas.

CHEMICAL BACKGROUND

Differentials can be exact or inexact. For an exact differential, the relationship

$$\left[\frac{\partial}{\partial y}\left(\frac{\partial f}{\partial x} \right)_y \right]_x = \left[\frac{\partial}{\partial x}\left(\frac{\partial f}{\partial y} \right)_x \right]_y$$

must hold. If this is the case, the function $f(x, y)$ is known as a state function. This means that its value is independent of the path taken to get to the particular state. This idea was discussed more fully in Section 3.3 on page 60.

Solution to Worked Example 3.17

We first need to write the pressure p of the ideal gas as a function of the other variables by rearranging the ideal gas equation. Since $pV = nRT$ we have

$$p = \frac{nRT}{V}$$

Since the variables on the right of this equation are the temperature T and the volume V, it may be clearer to write this using our notation for a function of two variables:

$$p(T,V) = \frac{nRT}{V}$$

To determine the differential dp, we need to calculate the partial derivatives

$$\left(\frac{\partial p}{\partial T}\right)_V \text{ and } \left(\frac{\partial p}{\partial V}\right)_T$$

These were determined in Worked Example 3.16, and were

$$\left(\frac{\partial p}{\partial T}\right)_V = \frac{nR}{V} \text{ and } \left(\frac{\partial p}{\partial V}\right)_T = -\frac{nRT}{V^2}$$

The differential dp will be given by the expression

$$dp = \left(\frac{\partial p}{\partial T}\right)_V dT + \left(\frac{\partial p}{\partial V}\right)_T dV$$

so we can substitute directly and obtain

$$dp = \left(\frac{nR}{V}\right)dT - \left(\frac{nRT}{V^2}\right)dV$$

It is relatively straightforward to show that

$$\left[\frac{\partial}{\partial V}\left(\frac{\partial p}{\partial T}\right)_V\right]_T = \left[\frac{\partial}{\partial T}\left(\frac{\partial p}{\partial V}\right)_T\right]_V = -\frac{nR}{V^2}$$

confirming that dp is an exact differential.

3.8 THE VAN DER WAALS EQUATION

We have already met the van der Waals equation which describes the behaviour of a gas in Worked Example 3.13, where we saw that corrections were made to both the pressure and volume to improve the description of real gases. These are the constants a and b in the equation

$$\left(p + \frac{an^2}{V^2}\right)(V - nb) = nRT$$

in which all the other symbols have the same meaning as in the ideal gas equation.

3.8.1 EXPANSION OF BRACKETS

The maths involved in the expansion of expressions using brackets has been discussed in Section 3.4.1 on page 64. We saw there that

$$(a+b)(c+d)=ac+ad+bc+bd$$

Worked Example 3.18

Remove the brackets from the van der Waals equation

$$\left(p+\frac{an^2}{V^2}\right)(V-nb)=nRT$$

which was introduced in Worked Example 3.13.

CHEMICAL BACKGROUND

The constants a and b in the van der Waals equation are normally obtained by fitting the equation to experimental measurements of pressure, volume and temperature.

Alternatively, since there is a critical temperature T_c above which a gas cannot be liquefied by pressure alone, it is possible to define a corresponding critical pressure p_c, and critical volume V_c. The quantities are related to the van der Waals constants by the equations

$$a=3p_cV_c^2 \text{ and } b=\frac{V_c}{3}$$

Solution to Worked Example 3.18

Multiplying the second bracket by each of the terms in the first bracket gives

$$p(V-nb)+\left(\frac{an^2}{V^2}\right)(V-nb)=nRT$$

Each term inside the remaining brackets of $(V-nb)$ can now be multiplied by the term outside, to give

$$pV-pnb+\left(\frac{an^2}{V^2}\right)V-\left(\frac{an^2}{V^2}\right)nb=nRT$$

We can now simplify the third and fourth terms on the left-hand side of this equation:

$$\left(\frac{an^2}{V^2}\right)V=\frac{an^2V}{V^2}=\frac{an^2}{V} \text{ and } \left(\frac{an^2}{V^2}\right)nb=\frac{an^3b}{V^2}$$

Since, using the rules of indices from Section 1.6,

$$\frac{V}{V^2} = V^{1-2} = V^{-1} = \frac{1}{V} \text{ and } n^2 \times n = n^{2+1} = n^3$$

so that the overall equation becomes

$$pV - pnb + \frac{an^2}{V} - \frac{an^3b}{V^2} = nRT$$

This would be a reasonable way of leaving the equation, as the brackets have been removed and all the terms simplified. However, a more elegant solution would be to continue as follows. The nRT term on the right can be 'removed' by subtracting nRT from both sides of the equation to give

$$pV - pnb + \frac{an^2}{V} - \frac{an^3b}{V^2} - nRT = 0$$

If we now multiply each term by V^2 (noting that $0 \times V^2 = 0$) we obtain

$$pV^3 - pnbV^2 + an^2V - an^3b - nRTV^2 = 0$$

or

$$pV^3 - (pnb + nRT)V^2 + an^2V - an^3b = 0$$

which can be classified as a cubic expression in V (see Section 3.4.2 on page 65) and contains no fractions.

3.8.2 COMBINING LIMITS

The concept of a limit was discussed in Section 3.5.1 on page 71. Limits can be combined in much the same way as the quantities they represent. For example, if

$$f(x) = g(x) + h(x)$$

then

$$\underset{x \to L}{\text{Lim}} f(x) = \underset{x \to L}{\text{Lim}} g(x) + \underset{x \to L}{\text{Lim}} h(x)$$

Similarly, if

$$f(x) = g(x)\, h(x)$$

then

$$\underset{x \to L}{\text{Lim}} f(x) = \underset{x \to L}{\text{Lim}} g(x) \underset{x \to L}{\text{Lim}} h(x)$$

Worked Example 3.19

Show that the van der Waals equation

$$\left(p+\frac{an^2}{V^2}\right)(V-nb)=nRT$$

reduces to the ideal gas equation in the limit of low pressure. See Worked Example 3.13 on page 93 for a definition of the terms in the equation.

CHEMICAL BACKGROUND

A general requirement of all acceptable equations of state for gases is that they will reduce to the ideal gas equation as the pressure falls to zero. The interactions between molecules are then negligible, as is the volume of the molecules relative to the volume of the gas, so we are approaching the ideal situation.

Solution to Worked Example 3.19

In terms of limits, we can write the van der Waals equation as

$$\underset{p\to 0}{\mathrm{Lim}}\left(p+\frac{an^2}{V^2}\right)\underset{p\to 0}{\mathrm{Lim}}(V-nb)=\underset{p\to 0}{\mathrm{Lim}}nRT$$

The first term to consider is

$$p+\frac{an^2}{V^2}$$

in the first bracket. We need to realize that, as the pressure p tends towards zero, volume V tends towards infinity. This can be expressed mathematically as $p \to 0$, $V \to \infty$.

As V becomes very large, $\frac{an^2}{V^2}$ will tend towards zero. Also because the V term is squared, this will happen faster than p tends to zero. Thus this term will tend to $p+0$, or just p. Mathematically we can express the steps in this argument as

$$\underset{p\to 0}{Lim}\left(p+\frac{an^2}{V^2}\right)=\underset{V\to\infty}{Lim}\left(p+\frac{an^2}{V^2}\right)=p$$

Now we consider the term in the second bracket. Again, as $p \to 0$, $V \to \infty$

so

$$\underset{p\to 0}{Lim}(V-nb)=\underset{V\to\infty}{Lim}(V-nb)$$

As V becomes very large, the quantity nb remains the same, and so the value of the second limit becomes closer to V:

$$\lim_{p \to 0} (V - nb) = V$$

Finally, as neither p nor V appears on the right of the equation, we have

$$\lim_{p \to 0} (nRT) = nRT$$

Substituting into our equation involving limits, this now gives us

$$pV = nRT$$

which is the ideal gas equation, as required.

Some of the arguments used above for determining each individual limit may seem difficult to replicate in different situations. Limits generally arise in the derivation of equations, so it is usually sufficient to be able to follow these arguments rather than developing them from scratch.

3.9 EQUILIBRIUM CONSTANTS

The equilibrium constant is a fundamental quantity in thermodynamics, describing the equilibrium state of a system in terms of concentration units. Its magnitude determines the extent to which a particular reaction will take place; a large value indicates a greater concentration of products than reactants and hence a favourable forward reaction. When defining the equilibrium constant we need to use a quantity known as the activity; this can be thought of as being a form of "effective concentration".

A mathematical way of defining the equilibrium constant K is to use the equation

$$K = \prod_j a_j^{v_j}$$

The symbol a_j here is used to represent the activity of species j; activities are the correct way of defining dimensionless equilibrium constants, i.e. those without units. The activity a_j of species j having concentration c_j can thus be defined by

$$a_j = \left(\frac{c_j}{c^\ominus} \right) \gamma_j$$

where $c^\ominus$ is the standard concentration, taken as 1 mol dm^{-3} exactly and γ_j is the activity coefficient of species j.

The mathematical symbol Π is new to us. It simply means 'multiply', or 'take the product'. The symbol v_j is known as the stoichiometric number of species j. This is obtained by rewriting our general chemical reaction

$$aA + bB + ... \rightarrow cC + dD + ...$$

as

$$cC + dD + aA - bB - 0$$

where the reactants have been subtracted from the products. We then have

stoichiometric number of $A = v_A = -a$
stoichiometric number of $B = v_B = -b$
stoichiometric number of $C = v_C = c$
stoichiometric number of $D = v_D = d$

Worked Example 3.20

Calculate the equilibrium constant for the dissociation of ethanoic acid CH_3COOH at 25°C if the equilibrium concentrations are

$$[CH_3COOH] = 0.90 \, mol \, dm^{-3}$$

and

$$[H^+] = [CH_3COO^-] = 0.004 \, mol \, dm^{-3}$$

FIGURE 3.15   Structure of ethanoic acid.

CHEMICAL BACKGROUND

Ethanoic acid (Figure 3.15) is more commonly known industrially by its older name of acetic acid. Along with acetic anhydride, it is produced by the catalytic oxidation of acetaldehyde and is used in the production of cellulose acetate which is needed for the manufacture of upholstery and carpets.

It is a weak electrolyte, and we can see from the concentration values above that the concentration of undissociated acid is far higher than that of the dissociated ions. We therefore expect to obtain an equilibrium constant value which is considerably less than 1. When the equilibrium constant is applied to acid dissociations such as this, it is usually given the symbol K_a. This leads to the use of logarithms for expressing acid strength in terms of pK_a, which is defined as

$$pK_a = -\log K_a$$

where log represents the logarithm taken to base 10, as we saw in Section 3.6.1 on page 80.

Solution to Worked Example 3.20

The first step in this problem is to write a balanced chemical equation for the dissociation of ethanoic acid. The species required are all given in the question and the required equation is

$$CH_3COOH \rightleftharpoons CH_3COO^- + H^+$$

Subtracting the reactant from the ethanoic acid product gives

$$CH_3COO^- + H^+ - CH_3COOH = 0$$

and the stoichiometric numbers are therefore

$$v_{CH_3COO^-} = 1$$

$$v_{H^+} = 1$$

$$v_{CH_3COOH} = -1$$

Substituting into the general expression for the equilibrium constant then gives

$$K = \left(\frac{c_{CH_3COO^-}}{c^\ominus} \gamma_{CH_3COO^-} \right)^1 \left(\frac{c_{H^+}}{c^\ominus} \gamma_{H^+} \right)^1 \left(\frac{c_{CH_3COOH}}{c^\ominus} \gamma_{CH_3COOH} \right)^{-1}$$

where c denotes the concentration of each species. As in this case, we often assume that the activity coefficients are all close to unity so that

$$K = \left(\frac{c_{CH_3COO^-}}{c^\ominus} \right)^1 \left(\frac{c_{H^+}}{c^\ominus} \right)^1 \left(\frac{c_{CH_3COOH}}{c^\ominus} \right)^{-1}$$

Since

$$\left(\frac{c_{CH_3COOH}}{c^\ominus} \right)^{-1} = \frac{1}{c_{CH_3COOH}/c^\ominus}$$

this can be rewritten as

$$K = \frac{\left(\dfrac{c_{CH_3COO^-}}{c^\ominus} \right) \left(\dfrac{c_{H^+}}{c^\ominus} \right)}{\left(\dfrac{c_{CH_3COOH}}{c^\ominus} \right)}$$

The values given in the question can now be substituted into this expression:

$$K = \frac{0.004 \times 0.004}{0.90} = 1.8 \times 10^{-5}$$

Notice that the use of $c^{\ominus}$ means that the units of mol dm^{-3} do not need to appear in the equation for calculating K.

3.9.1 SOLVING QUADRATIC EQUATIONS USING A FORMULA

In Section 3.4.2 on page 65, we saw that a polynomial expression of order 2 is called a quadratic. A quadratic equation is one of the form

$$ax^2 + bx + c = 0$$

where a, b and c are constants. Examples are

$$3x^2 + 2x - 1 = 0$$

and

$$x^2 + 6x + 4 = 0$$

The graphs of these equations are shown in Figure 3.16, where the solutions are given by the intersection of the curves with the x-axis, i.e. when the value of the expression is zero.

This may happen at none, one or two values of x. In some cases, these can be found by factorizing the expression into two brackets and setting each in turn to zero. However, this is not always possible, particularly for the problems we will meet in chemistry. Another method, which is more generally applicable, is to apply the standard formula.

The solution of the general quadratic equation given above is

$$x = \frac{-b \pm \sqrt{b^2 - 4ac}}{2a}$$

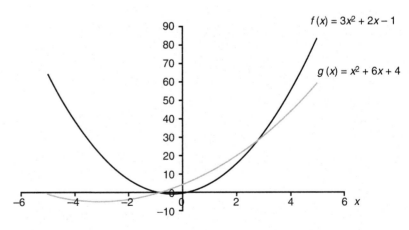

FIGURE 3.16   Graphs of the quadratic functions $f(x) = x^2 + 2x - 1$ and $g(x) = x^2 + 6x + 4$.

Notice that the±sign ensures that two values of x will result; however, solutions will only be found if b^2 is greater than the quantity $4ac$, since otherwise we would be taking the square root of a negative number. We will return to this topic again in Chapter 6.

Worked Example 3.21

In the reaction between nitrogen and hydrogen to give ammonia, the fractional amount of nitrogen reacted x can be related to the equilibrium constant K which has the value 977. This gives the quadratic equation

$$81.2x^2 - 163.4x + 81.2 = 0$$

for x. What is the value of x?

CHEMICAL BACKGROUND

Ammonia is manufactured industrially by means of the Haber process. This involves the reaction

$$N_{2(g)} + 3H_{2(g)} \rightleftharpoons 2NH_{3(g)}$$

which takes place at 450°C and 200 atmospheres in the presence of an iron catalyst. These conditions are used to give an acceptable reaction rate; the equilibrium constant gives us no information about this. Nitrogen is obtained from the air while hydrogen is usually obtained from the reaction of natural gas with steam. The Haber process is actually a major use for hydrogen gas.

Ammonia is also obtained commercially as a by-product of the generation of coke from coal. It is contained in an aqueous phase and is liberated by means of steam distillation. The largest use of ammonia is in fertilizers, but it is also used on a large scale in the production of nitric acid.

Solution to Worked Example 3.21

Relating the above expression to the general equation

$$ax^2 + bx + c = 0$$

gives $a = 81.2$, $b = -163.4$ and $c = 81.2$. Substituting these values into the equation for x

$$x = \frac{-b \pm \sqrt{b^2 - 4ac}}{2a}$$

gives

$$x = \frac{-(-163.4) \pm \sqrt{(-163.4)^2 - 4 \times 81.2 \times 81.2)}}{2 \times 81.2}$$

This can be solved in stages using a calculator to give

$$x = \frac{163.4 \pm \sqrt{26\,699.6 - 26\,373.8}}{162.4}$$

$$= \frac{163.4 \pm \sqrt{325.8}}{162.4}$$

$$= \frac{163.4 \pm 18.05}{162.4}$$

so

$$x = 1.12 \text{ or } x = 0.895$$

We have, as expected, obtained two values for x. However, if we look back at the question we see that only one is required. How do we know which one to take? If we look closely, we see that x is defined as the fractional amount of nitrogen reacted. Its value must therefore be less than 1 so we take $x = 0.895$ as our solution.

It is often necessary to eliminate one of the mathematical solutions in this way. Sometimes a negative and a positive value will be obtained, with the negative value obviously not making sense.

It is always worth checking your answers to these problems by substituting back into the given quadratic equation. Do remember, however, that due to rounding errors the answer that you have calculated may not give a value of exactly zero.

3.9.2 SOLVING QUADRATIC EQUATIONS ITERATIVELY

Although the equation introduced in the previous section is tedious to apply, it will give the solution to a quadratic equation without the need to understand any difficult mathematical concepts. However, it is possible to obtain the desired solution without using this formula.

Begin by remembering that x is a positive fractional amount, so its value must lie between 0 and 1. Using the function notation with

$$f(x) = 81.2x^2 - 163.4x + 81.2$$

we have

$$f(0) = 81.2 \times 0^2 - 163.4 \times 0 + 81.2 = 81.2$$

$$f(0.5) = 81.2 \times 0.5^2 - 163.4 \times 0.5 + 81.2 = 19.8$$

$$f(1.0) = 81.2 \times 1.0^2 - 163.4 \times 1.0 + 81.2 = -1.00$$

so that our solution must lie between $x = 0.5$ and $x = 1.0$ since $f(x) = 0$ represents the point where the curve crosses the x-axis.

In a similar fashion we determine

$$f(0.75)=81.2\times0.75^2-163.4\times0.75+81.2=4.33.$$

This is positive so our solution lies between $x=0.75$ and $x=1.0$. The half-way point between these is at $x=0.875$, so we calculate

$$f(0.875)=81.2\times0.875^2-163.4\times0.875+81.2=0.39$$

This is still positive, so our solution lies between $x=0.875$ and $x=1.0$. The midpoint is at $x=0.937\ 5$, so

$$f(0.937\ 5)=81.2\times0.937\ 5^2-163.4\times0.937\ 5+81.2=-0.62$$

Since this is negative, our solution lies between $x=0.875$ and $x=0.937\ 5$. The midpoint of these is $x=0.906$.

$$f(0.906)=81.2\times0.906^2-163.4\times0.906+81.2=-0.19$$

giving a root between $x=0.875$ and $x=0.906$. We once again determine the midpoint of these, which is $x=0.891$.

$$f(0.891)=81.2\times0.891^2-163.4\times0.891+81.2=0.07$$

so the root is between $x=0.891$ and $x=0.906$. We determine $f(x)$ at the midpoint, $x=0.899$

$$f(0.899)=81.2\times0.899^2-163.4\times0.899+81.2=-0.07$$

indicating a root between $x=0.891$ and $x=0.899$. The function at the midpoint $x=0.895$ is

$$f(x)=81.2\times0.895^2-163.4\times0.895+81.2=0.00$$

The root is therefore taken as $x=0.895$, as expected from our previous solution of this problem.

Note that this procedure only needs to be continued until the two values of x agree to within the desired degree of precision.

EXERCISES

1. Rearrange the following equations to make y the subject:
   a. $x+y=7$
   b. $9x-3y=18$
   c. $xy=12$
   d. $\dfrac{x}{2y}=6$

2. Rearrange the following equations to make y the subject:

   a. $\dfrac{1}{x} - \dfrac{1}{y} = 1$

   b. $4xy^2 = 16$

   c. $10 - 2xy = 0$

   d. $\dfrac{1}{x^2} - \dfrac{1}{2y} = 5$

3. Remove the brackets from the following expressions:

   a. $xy(x^2 + y^2)$

   b. $x(x+y)(x-y^2)$

   c. $(x+y)(x-y)$

   d. $x^2(x+a)(y-b)$

4. For each of the following functions:

   i. write down the order of the expression;

   ii. calculate $f(-3)$; and

   iii. write down the derivative $\dfrac{df(x)}{dx}$.

   a. $f(x) = 3x^4 + 2x^3 + 4x^2 + x + 6$

   b. $f(x) = 8x^6 + x^5$

   c. $f(x) = 8x^3 + 1$

   d. $3x^2 - 2x - 1$

5. Rewrite the expression for

$$f(x) = \frac{1}{x} - \frac{2}{x^2} + \frac{3}{x^3} - \frac{4}{x^4} + \frac{5}{x^5}$$

in terms of powers of x, and hence obtain an expression for $\dfrac{df(x)}{dx}$. What is the value of $\dfrac{df(x)}{dx}$ when $x = 1.1$?

6. Give the values of the following quantities:

   a. $\log 24$

   b. $\log 10$

   c. $\ln 3$

   d. $\ln 2.718$

7. Give the gradients and intercepts of the straight lines represented by the following equations:

   a. $y = 6x + 3$

   b. $3y = 5x + 4$

   c. $x - 2y = 8$

   d. $\dfrac{y}{x} = 4$

8. For the following data plot $\ln y$ against $\dfrac{1}{x}$ using a pencil and graph paper:

x	1.0	2.0	3.0	4.0	5.0
y	0.368	0.135	0.097	0.082	0.074

and determine the gradient of the resulting graph.

9. For the following data plot $\dfrac{1}{y}$ against x^2 using a spreadsheet graphing function:

x	0.0	1.0	2.0	3.0	4.0
y	−2.00	0.400	0.086 9	0.037 7	0.021 6

What is the gradient of the resulting graph?

10. If x and y are directly proportional, and x takes the value 10 when y is 30, what is the value of the proportionality constant?

11. If x and y are inversely proportional, and x takes the value 6 when y is 3, what is the value of x when y is 10?

12. State the relationship between x and y if they can take the following pairs of values:

x	64	32	16	8	4	2	1
y	1	2	4	8	16	32	64

13. If $f(x, y) = 3x^2y - 2xy$ calculate
   a. $f(1, -1)$
   b. $f(0, 2)$
   c. $f(-2, 1)$
   d. $f(-2, -2)$

14. If $f(x, y) = \dfrac{1}{x} - \dfrac{1}{y}$ calculate
   a. $f(-1, -1)$
   b. $f(1, 1)$
   c. $f(1, -2)$
   d. $f(-2, 1)$
   When is $f(x, y)$ not defined?

15. If $f(x, y)=xy+\ln(xy)$ calculate
    a. $f(1, 1)$
    b. $f(1, 2)$
    c. $f(1, 3)$
    d. $f(0.5, 0.75)$
    For what values of x and y is $f(x, y)$ not defined?

16. If $f(x, y)=3x^2+2xy+4x^3y^2$ calculate

$$\frac{\partial f}{\partial x} \text{ and } \frac{\partial f}{\partial y}$$

    and write an expression for the differential $df(x, y)$.

17. Show that the $df(x, y)$ in Question 16 is an exact differential.

18. If $f(x, y)=x^2+x^2y+xy^2+y^2$ calculate

$$\frac{\partial f}{\partial x} \text{ and } \frac{\partial f}{\partial y}$$

19. Solve these quadratic equations using the formula:
    a. $x^2+9x-7=0$
    b. $4x^2+7x+2=0$
    c. $0.20x^2-0.90x+0.20=0$
    d. $0.5x^2-2.2\,x-4.7=0$

20. The quadratic equation

$$0.24x^2+0.20x-0.16=0$$

    has one root between $x=-1.5$ and $x=-1.0$, and another between $x=0.3$ and $x=0.6$. Find these roots using an iterative trial and error method.

PROBLEMS

1. Rearrange each of the following equations to make the given variable the subject:
    a. $\Delta G=\Delta H - T\Delta S$ (subject T)
    b. $V=n_1V_1+n_2V_2$ (subject V_1)
    c. $F=C - P+2$
    d. $P_1V_1=p_2V_2$

2. If a gas obeys the ideal gas equation

$$pV=nRT$$

state the value of the gradient obtained when each of the following plots was made:
a. p against T at constant n and V
b. p against n at constant V and T
c. V against n at constant p and T.

3. A simplified representation of the heat capacity C_p in terms of
the absolute temperature T is given by

$$C_p=a+bT$$

where a and b are constants. If C_p is measured at various values
of T, how can the values of a and b be determined?

4. The equilibrium constant K is related to the standard enthalpy
of reaction $\Delta H^\ominus$ and the standard entropy of reaction $\Delta S^\ominus$ by the
equation

$$\ln K = -\left(\frac{\Delta H^\ominus}{RT}\right)+\left(\frac{\Delta S^\ominus}{R}\right)$$

where R is the ideal gas constant and T is the absolute temperature. Calculate

$$\frac{d(\ln K)}{dT}$$

assuming that $\Delta H^\ominus$ and $\Delta S^\ominus$ are independent of temperature.

5. We saw in the chapter that the heat capacity C_p at constant pressure can be represented by the equation

$$C_p = a+bT+\frac{c}{T^2}$$

Obtain an expression for $\dfrac{dC_p}{dT}$ in terms of a, b, c and T.

6. Molecular speeds can be measured by the root-mean-square
speed v. This is a function of molecular mass M and absolute
temperature T and can be written as

$$v(M,T)=\left(\frac{3RT}{M}\right)^{\frac{1}{2}}$$

Calculate the root-mean-square speed at 298 K of carbon dioxide
for which $M=44.0\,\text{g mol}^{-1}$.

7. Derive an expression for the differential dV for a gas which obeys the ideal gas equation

$$pV = nRT$$

assuming that n is a constant.

8. The van der Waals constants for oxygen used to be given as $a = 1.36$ atm l^2 mol^{-2} and $b = 0.031\ 8\ l$ mol^{-1}. Use the conversion factors below to express these quantities in the appropriate SI units, which are Pa m^{6}mol^{-2} and m^{3}mol^{-1} respectively.

$$1\ atm = 101.325\ kPa$$

$$1\ l = 1\ dm^3$$

$$1\ Pa = 1\ N\ m^{-2}$$

9. The virial equation of state can be written as a power series in pressure p as

$$pV = RT + Bp + Cp^2 + ...$$

where B and C are functions of temperature. By dividing this equation by T show that, in the limit of zero pressure, this equation reduces to the ideal gas equation.

10. The equilibrium constant K for the dissociation of hydrochloric acid is given by the formula

$$K = \frac{c\alpha^2}{1 - \alpha}$$

where c is the concentration of acid and α is the degree of dissociation. Rearrange this equation to give a quadratic equation in α. Solve this equation using the quadratic formula for an acid solution of concentration 0.05 mol dm^{-3} if $K = 0.40$.

11. At relatively low pressures, it is possible to ignore the constant b in the van der Waals equation to give the equation of state

$$pV^2 - RTV + a = 0$$

Solve this quadratic equation in V using the formula.

12. The position of equilibrium x in a mixture of N_2O_4 and NO_2 that initially consists of 1.000 mol of dinitrogen tetroxide at 298 K and confined to a volume of 25 dm³ is given by the equation

$$3.912\, x^2 + 0.148\,4\, x - 0.148\,4 = 0$$

Use an iterative method to determine the value of x that lies between 0.15 and 0.20.

13. Show that the equation

$$\Delta G = -nFE$$

is dimensionally correct given that ΔG is measured in J, n is a pure number and E is measured in V. The value of the Faraday constant F can be found in Appendix B.

14. Show that the equation for the elevation of the boiling point

$$\Delta_{vap}T = \frac{M_1 R T_b^{*2}}{\Delta_{vap}H} m_2$$

is dimensionally correct where $\Delta_{vap}T$ is measured in K, M_1 in g mol⁻¹, T_b^{*} in K, $\Delta_{vap}H$ in kJ mol⁻¹ and m² in mol kg⁻¹. R is the gas constant.

Solution Chemistry

4

4.1 INTRODUCTION

In some respects, the physical chemistry of liquid solutions is more complicated than that of either gases or solids. Molecules in the gas phase have relatively large separations, so the interactions between them are reduced and many approximations may be used to simplify the maths involved. Conversely, if a solid is crystalline it can be considered to be composed of many repeats of a simple unit, which again reduces the complexity of the maths required.

In this chapter, we will be concerned only with aqueous solutions, as these form by far the largest group of interest.

4.2 CONCENTRATION OF SOLUTIONS

4.2.1 CONCENTRATION OF A SOLUTION

The concentration of a solution is calculated from the simple equation

$$c = \frac{n}{V}$$

where n is the amount of substance dissolved in volume V. In the laboratory we are more interested in determining the mass m of a substance that needs to be dissolved, and this is related to the amount by the equation

$$n = \frac{m}{M}$$

where m is the molar mass, typically given in units of g mol^{-1}. If we combine these equations we have

$$c = \frac{n}{V} = \frac{m/M}{V} = \frac{m}{MV}$$

which gives us the concentration in terms of the mass dissolved.

DOI: 10.1201/9781003043218-4

Worked Example 4.1

a. What is the concentration of the resulting solution when 25.0 g of sodium chloride is dissolved in 100 cm³ of water?
b. What amount of sodium chloride is present in 300 cm³ of a solution of concentration 0.25 mol dm⁻³?

CHEMICAL BACKGROUND

Sodium chloride is very soluble in water because it is an ionic compound. Since there are two isotopes of chlorine, its atomic mass is a weighted value based on the abundance of each.

Solution to Worked Example 4.1

a. The first step in this problem is to calculate the molar mass of NaCl, which is simply the sum of the atomic masses of sodium and chlorine. This is

$$M_{NaCl} = M_{Na} + M_{Cl}$$

$$= 22.99 \text{ g mol}^{-1} + 35.45 \text{ g mol}^{-1}$$

$$= 58.44 \text{ g mol}^{-1}$$

We then need to consider the volume we have been given and convert this to units of dm³ to obtain a concentration value in mol dm⁻³. Since 1 dm = 10 cm it follows that 1 cm = 0.1 dm and so

$$100 \text{ cm}^3 = 100 \ (0.1 \text{ dm})^3 = 10^2 \times 10^{-3} \text{ dm}^3$$

Combining these using the rules of indices in Section 1.6.1 on page 17 gives

$$100 \text{ cm}^3 = 10^{-1} \text{ dm}^3 = 0.100 \text{ dm}^3$$

Substituting into the equation for concentration then gives

$$c = \frac{m}{MV} = \frac{25.0 \text{ g}}{58.44 \text{ g mol}^{-1} \times 0.100 \text{ dm}^3}$$

and a concentration of 4.28 mol dm⁻³.
b. We can convert the volume of 300 cm³ by using the same technique as before.

$$V = 300 \text{ cm}^3 = 300 \times (10^{-1} \text{ dm})^3 = 3 \times 10^2 \times 10^{-3} \text{ dm}^3 = 0.300 \text{ dm}^3$$

We then rearrange our first equation by multiplying both sides by V to give

$$n = cV$$

$$= 0.25 \text{ mol dm}^{-3} \times 0.300 \text{ dm}^3 = 0.075 \text{ mol}$$

as the units dm⁻³ and dm³ cancel.

4.2.2 DILUTION OF A SOLUTION

We saw in the previous section that

$$n=cV$$

If we have an initial volume V_1 of a solution of concentration c_1 and dilute it to a new volume V_2 its concentration will reduce to value c_2 but the amount of solute present will remain unchanged as n. We can therefore write

$$n=c_1V_1=c_2V_2$$

It is more useful to leave just the concentration and volume terms so we have

$$c_1V_1=c_2V_2$$

Worked Example 4.2

What volume of concentrated sulfuric acid at $18\,\text{mol dm}^{-3}$ needs to be diluted to obtain $1\,\text{dm}^3$ of a solution with a concentration of $0.25\,\text{mol dm}^{-3}$?

CHEMICAL BACKGROUND

Considerable heat is liberated in the dilution of sulfuric acid, so the concentrated acid should always be added to water rather than the other way round. It is manufactured in large quantities due to its use in the production of many materials, most production being via the contact process involving the following reactions:

$$S_{(s)}+O_{2(g)} \rightarrow SO_{2(g)}$$

$$2\,SO_{2(g)}+O_{2(g)} \rightarrow 2\,SO_{3(g)}$$

$$SO_{3(g)}+H_2O_{(l)} \rightarrow H_2SO_{4(aq)}$$

with the second reaction taking place on the surface of solid vanadium (V) oxide, V_2O_5.

Solution to Worked Example 4.2

The initial concentration c_1 is $18\,\text{mol dm}^{-3}$ and we wish to dilute this to concentration c_2 of $0.25\,\text{mol dm}^{-3}$. The final volume V_2 is $1\,\text{dm}^3$ and we wish to obtain the initial volume V_1 required. Rearranging the equation

$$c_1V_1 = c_2V_2$$

gives

$$V_1 = \frac{c_2V_2}{c_1} = \frac{0.25 \text{ mol dm}^{-3} \times 1 \text{ dm}^3}{18 \text{ mol dm}^{-3}} = 0.014 \text{ dm}^3$$

In practical terms, it is more useful to convert this to units of cm^3. Since $1\,dm = 10\,cm$, we have

$$V_1 = 0.014\,(10\,cm)^3 = 0.014 \times 10^3\,cm^3 = 14\,cm^3$$

4.3 ACTIVITY

We met the concept of activity briefly in Section 3.9 on page 109. For a solution of concentration c, the activity a is defined as

$$a = \left(\frac{c}{c^{\ominus}}\right)\gamma_{\pm}$$

where $c^{\ominus}$ is the standard concentration of exactly $1\,mol\,dm^{-3}$, and $\gamma_{\pm}$ is a constant known as the activity coefficient. In dilute solutions, $\gamma_{\pm}$ is close to 1 so activities are often taken as being equal to concentrations.

Worked Example 4.3

The activity coefficient of a $0.050\,mol\,dm^{-3}$ solution of NaCl in water is 0.820. Determine the activity of this solution.

CHEMICAL BACKGROUND

Activity coefficients can be determined using a variety of techniques, which include vapour pressure measurements, freezing point depression, solubility, and electrochemistry. The quantity $\gamma_{\pm}$ is strictly the mean activity coefficient; this is related to the activity coefficients γ_+ and γ_- for positive and negative ions respectively by

$$\gamma_{\pm} = \sqrt{\gamma_+\gamma_-}$$

Solution to Worked Example 4.3

We have all the values we require to substitute into the defining equation to give

$$a = \left(\frac{0.050\,mol\,dm^{-3}}{1\,mol\,dm^{-3}}\right) \times 0.820$$

The units cancel top and bottom to leave

$$a = \frac{0.05}{1} \times 0.820 = 0.04$$

Notice that we give the answer to 1 significant figure, as this is the minimum level of precision in the quantity $0.05\,mol\,dm^{-3}$ in the original data.

4.4 MOLALITY

We are familiar with solutions whose concentration is expressed in units of mol dm^{-3}. Concentration is accordingly defined as the amount of solute in 1 dm^3 of **solution**. An alternative way of expressing the strength of solution is by using its molality. This has units of mol kg^{-1}, and is defined as the amount of solute in 1 kg of **solvent**.

Note that the volume of a solution changes with temperature but the mass of solvent doesn't. Molality is thus independent of temperature while concentration isn't. It therefore appears in, for example, the expressions for calculating boiling point elevation and freezing point depression of solutions.

4.4.1 PROPORTION

Proportion was introduced in Section 3.7.4 on page 95. We saw there that two quantities were directly proportional if $\dfrac{y}{x}$ is constant, and inversely proportional if xy is constant.

Worked Example 4.4

0.50 mol of HCl was dissolved in 0.75 kg of water. What amount of HCl should be dissolved in 2.50 kg of water to give a solution of the same molality?

CHEMICAL BACKGROUND

Hydrogen chloride gas consists of covalently bonded HCl molecules with weak interactions between them. These readily dissolve in water to form separate H^+ and Cl^-; it is the former which gives the resulting solution its acidic properties.

Solution to Worked Example 4.4

A little thought should suggest that for a series of solutions having the same molality, the amount of HCl will be directly proportional to the mass of water. Consequently for the first solution, we have

$$\frac{n_1}{M_1} = \frac{0.50\,\text{mol}}{0.75\,\text{kg}} = 0.67\,\text{mol}\,\text{kg}^{-1}$$

where n_1 is the amount of HCl and M_1 the mass of water in the first solution.

In this problem, the constant $\dfrac{n_1}{M_1}$ is our constant of proportionality, but you may also realise from the units that it is in fact the molality of the solution. For the second solution we can therefore write

$$\frac{n_2}{2.50\,\text{kg}} = 0.67\,\text{mol}\,\text{kg}^{-1}$$

where n_2 is the amount of HCl in this second solution. If we multiply both sides of this equation by 2.50 kg we obtain

$$n_2 = 0.67 \, \text{mol kg}^{-1} \times 2.50 \, \text{kg} = 1.68 \, \text{mol}$$

4.5 RAOULT'S LAW

Raoult's Law allows us to calculate the vapour pressure above a solution which contains a mixture of two liquids. If this vapour pressure is p, then

$$p = x_1 p_1^* + x_2 p_2^*$$

where x_1 and x_2 are the mole fractions of components 1 and 2 respectively in the mixture, and p_1^* and p_2^* are the vapour pressures of the pure components 1 and 2 respectively. This relationship is illustrated graphically in Figure 4.1.

4.5.1 STRAIGHT LINE GRAPHS

We saw in Section 3.6.2 on page 83 that the equation of a straight line is given by

$$y = mx + c$$

where x and y are the variables, m is the gradient and c is the intercept, as shown in Figure 4.2.

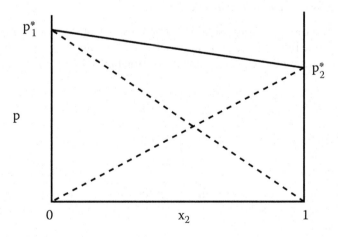

FIGURE 4.1   Graph of pressure against mole fraction illustrating Raoult's law.

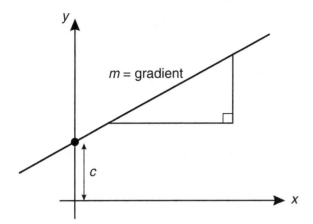

FIGURE 4.2   The straight-line graph.

Worked Example 4.5

Show that Raoult's law can be expressed as a straight line graph of p against the mole fraction x of one of the components.

CHEMICAL BACKGROUND

Raoult's law is observed most closely when the two components in the solution are quite similar, such as benzene and toluene for example. This is because in such a solution the interactions between benzene and benzene, toluene and toluene, and benzene and toluene will be similar.

Solution to Worked Example 4.5

The important conceptual step in this problem is the realisation that the mole fractions in a mixture must add to 1 so that

$$x_1 + x_2 = 1$$

We can rearrange this expression by subtracting x_2 from either side to give

$$x_1 + x_2 - x_2 = 1 - x_2$$

or

$$x_1 = 1 - x_2$$

We can now substitute this into the defining equation for Raoult's law

$$p = x_1 p_1^{\cdot} + x_2 p_2^{\cdot}$$

to give

$$p = x_1 p_1^{\bullet} + (1-x_1) p_2^{\bullet}$$

Removing the brackets (as in Section 3.4.1 on page 64) gives us

$$p = x_1 p_1^{\bullet} + p_2^{\bullet} - x_1 p_2^{\bullet}$$

The order of addition is unimportant so we can write this as

$$p = x_1 p_1^{\bullet} - x_1 p_2^{\bullet} + p_2^{\bullet}$$

Collecting the terms in x_1 now leads to

$$p = \left(p_1^{\bullet} - p_2^{\bullet} \right) x_1 + p_2^{\bullet}$$

which is in the form $y = mx + c$, the equation of a straight line, with variables x_1 and p, gradient $(p_1^{\bullet} - p_2^{\bullet})$ and intercept $p_2^{\bullet}$.

Notice that if we did take measurements of x_1 and p, $p_2^{\bullet}$ is given directly by the intercept and once we have its value we can find that of $p_1^{\bullet}$ from the gradient.

Finally, notice that this problem could also be worked through with x_2 as the x-axis variable, rather than x_1.

4.5.2 PROPORTION

Proportion was introduced in Section 3.7.4 on page 95. We saw there that two quantities were

- directly proportional if $\dfrac{y}{x}$ is constant, and

- inversely proportional if xy is constant.

Worked Example 4.6

Raoult's law can also be written as

$$p = p_1 + p_2$$

where p_1 and p_2 denote the contributions to the overall pressure of components 1 and 2. These are known as partial pressures and are given by

$$p_1 = x_1 p_1^{\bullet} \quad \text{and} \quad p_2 = x_2 p_2^{\bullet}$$

using the same terms as before.

Air at a temperature of 298 K and a pressure of 786 mm of mercury contains nitrogen at a partial pressure of 613 mm of mercury. Use this data to calculate the percentage of nitrogen in the air.

CHEMICAL BACKGROUND

In a mercury barometer, the height of mercury is proportional to the barometric pressure. This depends on the altitude and on weather conditions.

Solution to Worked Example 4.6

Dividing the first equation for partial pressure, $p_1 = x_1\, p_1^*$, by p_1^* gives

$$x_1 = \frac{p_1}{p_1^*}$$

Substituting values for p_1 and p_1^* gives

$$x_1 = \frac{613\,\text{mm Hg}}{786\,\text{mm Hg}} = 0.780$$

There is no need to convert the rather inconvenient units of mm Hg as they are the same on the top and bottom of this fraction and so cancel. Note also that we include the trailing zero in 0.780 to give three significant figures as this level of precision is given for each of the pressure values.

Since we were asked for the percentage value we have to multiply the fraction by 100 to give the familiar value of 78.0%.

4.6 THE DEBYE-HÜCKEL EQUATION

The Debye-Hückel Limiting Law expresses the mean activity coefficient $\gamma_\pm$, introduced in Section 4.3, in terms of the ionic strength I and the charges z_+ and z_- on the positive and negative ions respectively:

$$\log \gamma_\pm = -0.51 \left| z_+ z_- \right| \sqrt{\frac{I}{\text{mol dm}^{-3}}}$$

The symbol '$|x|$' means 'take the value of x without regard to sign'; this is called the **modulus** of x. Thus $|-5|$ would be 5, and $|5|$ would also be 5. The symbol 'log' is used to refer to logarithms to the base 10 (see Section 3.6.1), and the reason for dividing I by units of mol dm^{-3} is to ensure that the quantity on the right of the equation has no units. This must be true on the left of the equation as the quantity is logarithmic and the logarithm of a number has no units. As we saw in Section 4.3 on page 126 the mean activity coefficient $\gamma_\pm$ is defined as

$$\sqrt{\gamma_+ \gamma_-}$$

for a 1: 1 electrolyte, with γ_+ and γ_- being the activity coefficients of the positive and negative ions respectively.

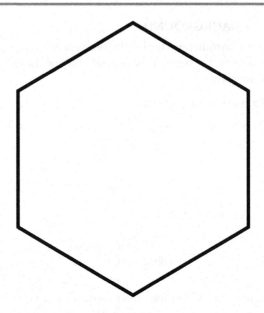

FIGURE 4.3   Structure of cyclohexane.

The Debye-Hückel Limiting Law gives reasonable results at low concentrations, but the deviations from it as concentrations increase are considerable. The reduction in the activity coefficient of an ion is due to its interactions with the surrounding ionic atmosphere; the effect increases with charge and in solvents of low dielectric constant, as both these factors increase the magnitude of the interactions.

The relative dielectric constant is defined as the ratio of the dielectric constant of the medium to the dielectric constant of a vacuum. Consequently, it has no units. Typical values of the relative dielectric constant are 78.54 for water and 2.015 for cyclohexane (Figure 4.3) at 25°C.

4.6.1 LOGARITHMS

The subject of logarithms was discussed in Section 3.6.1 on page 80. Here we saw that the term 'log' is used to refer to logarithms to base 10, and that if

$$A = 10^c$$

then

$$c = \log_{10} a$$

or more usually

$$c = \log a$$

Worked Example 4.7

Use the Debye-Hückel Limiting Law to calculate the ionic strength of a solution of potassium chloride having a mean activity coefficient of 0.927.

CHEMICAL BACKGROUND

The ionic strength I of a solution is defined by the equation

$$I = 0.5 \sum_i m_i z_i^2$$

where m_i is the molality of the ions i in the solution, and z_i is their charge. Notice that this expression makes use of the sigma notation Σ, introduced in Section 2.4.3 on page 44, which simply means take the sum. In this case, the sum relates to all components of the solution.

In dilute solutions, it is usually justifiable to replace the molalities m_i by concentrations c_i. The expression then becomes

$$I = 0.5 \sum_i c_i z_i^2$$

The rates of ionic reactions are found to vary between solutions of different ionic strength, and the second-order rate constant k is given by the equation

$$\log k = \log k_o + 1.02 \, z_A \, z_B \sqrt{\frac{I}{\mathrm{mol\,dm}^{-3}}}$$

where k_o is a constant which depends upon the concentrations of the activated complexes present.

Solution to Worked Example 4.7

The first stage in this problem is to rearrange the equation which expresses the Debye-Hückel Limiting Law so that the term in ionic strength appears on its own on the left-hand side. Dividing both sides by $-0.51 \, |z_+ z_-|$ and reversing the equation gives us the expression

$$\sqrt{\frac{I}{\mathrm{mol\,dm}^{-3}}} = \frac{\log \gamma_\pm}{-0.51 |z_+ z_-|}$$

In this case, it is probably easier to obtain a value for

$$\sqrt{\frac{I}{\mathrm{mol\,dm}^{-3}}}$$

and then square it, rather than trying to calculate I directly. Since, in solution, potassium chloride consists of K^+ and Cl^- ions, the values of the charges z_+ and z_- are 1 and -1 respectively. Substituting in the rearranged equation above now leads to

$$\sqrt{\frac{I}{\text{mol dm}^{-3}}} = \frac{\log 0.927}{-0.51 \times |1 \times (-1)|}$$

Since the modulus $|1 \times (-1)| = |-1| = 1$, we then have

$$\sqrt{\frac{I}{\text{mol dm}^{-3}}} = \frac{-\log 0.927}{0.51}$$

$$= \frac{-(-0.032\,9)}{0.51}$$

$$= 0.064\,5$$

Note that the value of $\log 0.927$ is negative. It is a general result that the logarithm of a fractional number will be negative.

To obtain the ionic strength, we now have to square both sides to give

$$\frac{I}{\text{mol dm}^{-3}} = 0.064\,5^2 = 0.004\,2$$

and therefore $I = 0.004\,2\,\text{mol dm}^{-3}$.

4.7 OSTWALD'S DILUTION LAW

The molar conductivity Λ of an ionic solution is a measure of how well it conducts electricity and is measured in units of $\Omega^{-1}\text{cm}^2\text{mol}^{-1}$ or S $\text{cm}^2\text{mol}^{-1}$. ($\Omega$ is the symbol for the ohm, the unit of electrical resistance and S is the symbol for siemens where $1\,S = 1\,\Omega^{-1}$.) It is possible to extrapolate measured values of Λ to zero concentration and obtain a value for Λ_0, the molar conductivity at infinite dilution.

Examples of molar conductivities are those for K^+ and Cl^- which are 73.5 and 76.4 S $\text{cm}^2\text{mol}^{-1}$ respectively, at a concentration of 0.01 mol dm^{-3}. The overall conductivity of a solution is found by multiplying the molar conductivity by concentration and adding these products for all those ions present.

For the dissociation of an electrolyte XY according to the equation

$$XY \rightleftharpoons X^+ + Y^-$$

the equilibrium constant K_c, in terms of concentration, is defined as

$$K_c = \frac{[X^+][Y^-]}{[XY]}$$

but it is also possible to define the dimensionless equilibrium constant K in terms of the molar conductivities:

$$K = \frac{c\left(\dfrac{\Lambda}{\Lambda_o}\right)^2}{1-\left(\dfrac{\Lambda}{\Lambda_o}\right)}$$

which is a satisfactory expression for many weak electrolytes.

4.7.1 DISCONTINUITIES

A function has a **discontinuity** when it is not defined for every possible input value. For example, consider the function defined by

$$f(x) = \frac{x+1}{x-4}$$

For a function to be defined, there has to be a single value that results from evaluating the function for a given value of x. If we evaluate the function f for the first few integers, we obtain

$$f(0) = \frac{1}{-4} = -\frac{1}{4}$$

$$f(1) = \frac{2}{-3} = -\frac{2}{3}$$

$$f(2) = \frac{3}{-2} = -\frac{3}{2}$$

$$f(3) = \frac{4}{-1} = -4$$

$$f(4) = \frac{5}{0}$$

When $x=4$, the denominator of $f(x)$ is zero. Division by zero results in an infinite number of possible answers, none of which are acceptable, so $f(x)$ is not defined when $x=4$. A plot of the function $f(x)$ is shown in Figure 4.4. Notice that if the value of x is very close to 4 then the function is still defined; it is discontinuous only at the exact value $x=4$.

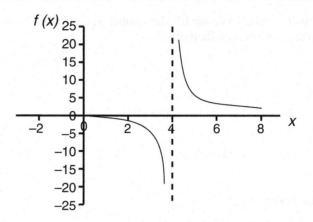

FIGURE 4.4   Graph of $f(x) = \dfrac{x+1}{x-4}$.

Worked Example 4.8

The expression of Ostwald's Dilution Law above is an example of a function which has a discontinuity. In this case, it is not possible to obtain a value of K for every value of Λ. For what value or values of Λ is K undefined?

CHEMICAL BACKGROUND

For an electrolyte MX, the degree of dissociation α is defined as

$$\alpha = \frac{m_{X^-}}{m}$$

where m_{X^-} is the molality of the dissociated anion and m is the overall molality of the MX solution.

For strong electrolytes, the concept of degree of dissociation does not apply. If they are virtually fully ionized the degree of dissociation a will be equal to 1, and the properties of their solutions will depend to a large extent on the nature of the solvent.

Solution to Worked Example 4.8

In the equation

$$K = \frac{c\left(\dfrac{\Lambda}{\Lambda_0}\right)^2}{1 - \left(\dfrac{\Lambda}{\Lambda_0}\right)}$$

K will be undefined when the denominator is zero, i.e. when

$$1 - \frac{\Lambda}{\Lambda_0} = 0$$

Adding $\dfrac{\varLambda}{\varLambda_o}$ to both sides of this equation gives

$$\frac{\varLambda}{\varLambda_o} = 1$$

It follows that K is not defined when $\varLambda = \varLambda_o$.

4.8 PARTIAL MOLAR VOLUMES

When two solutions are mixed, the total volume of the resulting solution is not necessarily the sum of the individual volumes. The total volume V of solution can be expressed by the equation

$$V = n_1 V_1 + n_2 V_2$$

where n_1 and n_2 are the amounts (measured in mol) of components 1 and 2, and V_1 and V_2 are the corresponding respective partial molar volumes. The partial molar volume is actually defined as a partial derivative:

$$V_1 = \left(\frac{\partial V}{\partial n_1} \right)_{T, p, n_2, \dots}$$

The subscripts T, p, n_2,... indicate that the temperature, pressure and amount of component 2 are all held constant as we consider the variation of volume V with the amount n_1 of component 1. The dots allow for solutions composed of more than two components and represent n_3, n_4, etc.

Typical values of partial molar volumes are 16.98 and 57.60 cm^{3}mol^{-1} for water and ethanol respectively in a solution containing 0.42 mole fraction of ethanol.

4.8.1 FUNCTIONS

The subject of functions was discussed in Section 3.4.3 on page 69. The value of a function $f(x)$ is found by substituting a specific value of x into the expression everywhere that the symbol for x appears.

Worked Example 4.9

The partial molar volume V_1 of water in a solution of potassium sulfate is given as a function of molality m by the equation

$$\frac{V_1}{\text{cm}^3} = -0.109\ 4 \left(\frac{m}{m^{\ominus}} \right)^{\frac{3}{2}} - 0.000\ 2 \left(\frac{m}{m^{\ominus}} \right)^2 + 17.963$$

where $m^{\ominus}$ is the standard molality of 1 mol kg^{-1}.

Calculate the partial molar volume of water when m $= 0.10$ mol kg^{-1}.

CHEMICAL BACKGROUND

Although partial molar volumes are defined in terms of the amounts n_1 and n_2 of solvent and solute, for practical purposes we often work in terms of the molality m of the solute. As we saw in Section 4.4, this is defined as the amount of solute per kg of solvent, so if we actually have (or can assume) 1 kg of solvent then m will be the same as n_2. The conversion from concentration to molality can be made if we know the density of the solution.

The inclusion of units in the expression above may seem cumbersome, but it does ensure that both sides of the equation can be evaluated to a pure number (without units) so the equation is valid both numerically and in terms of units.

Solution to Worked Example 4.9

The expression for V_1 can be written using the function notation as

$$\left(\frac{V_1}{\text{cm}^3}\right)\left(\frac{m}{m^{\ominus}}\right) = -0.109\,4\left(\frac{m}{m^{\ominus}}\right)^{\frac{3}{2}} - 0.000\,2\left(\frac{m}{m^{\ominus}}\right)^2 + 17.963$$

and its graph is shown in Figure 4.5. The expression above appears cumbersome at first sight, but inclusion of the molality term on the left of the equals sign does emphasize that this is the variable in the function V_1. We can calculate V_1 for $0.10\,\text{mol kg}^{-1}$ by direct substitution into the function expression. This gives us

$$\left(\frac{V_1}{\text{cm}^3}\right)\left(\frac{0.10\,\text{mol kg}^{-1}}{1\,\text{mol kg}^{-1}}\right)$$

$$= -0.109\,4\left(\frac{0.10\,\text{mol kg}^{-1}}{1\,\text{mol kg}^{-1}}\right)^{\frac{3}{2}} - 0.000\,2\left(\frac{0.10\,\text{mol kg}^{-1}}{1\,\text{mol kg}^{-1}}\right)^2 + 17.963$$

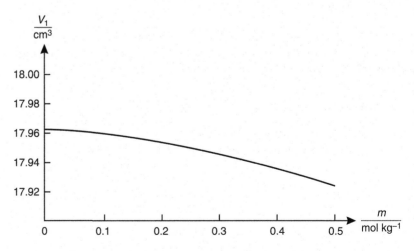

FIGURE 4.5   Graph of the function describing the partial molar volume of water in a solution of potassium sulfate.

and subsequently

$$\left(\frac{V_1}{cm^3}\right)(0.1) = -0.109\,4 \times 0.10^{\frac{3}{2}} - 0.000\,2 \times 0.10^2 + 17.963$$

Evaluating this expression term by term using a calculator gives us

$$\frac{V_1(0.1)}{cm^3} = -0.109\,4 \times 0.031\,62 - 2 \times 10^{-6} + 17.963$$

$$= -0.003\,46 - (2 \times 10^{-6}) + 17.963$$

The second term of 2×10^{-6} is much smaller than the other two and can be safely neglected because the final term is given to 3 decimal places. We are now left with

$$\frac{V_1(0.10)}{cm^3} = -0.003\,46 + 17.963$$

$$= 17.959\,54$$

We are therefore restricted to three decimal places in our answer which will be $17.960\,cm^3$. Thus we can write $V_1\,(0.10) = 17.960\,cm^3$.

4.8.2 STATIONARY POINTS

We saw in Section 3.5.1 on page 71 that we could think of the derivative $\frac{dy}{dx}$ as being the gradient of the graph of y against x, which could also be found by drawing the tangent to the curve at the point of interest. Suppose that we have a function $f(x)$ whose graph is shown in Figure 4.6. This has a maximum (peak) and a minimum (trough) at the points indicated; these are known as **stationary points.** Notice that at both the maximum and the minimum the tangent to the curve is horizontal, and consequently its gradient has a value of zero.

This fact gives us a way of determining the points at which a function $f(x)$ has a stationary point. If the tangent to the curve is zero, then

$$\frac{df(x)}{dx} = 0$$

It is also possible to identify the nature of the stationary points found. This is done by looking at the value of the second derivative $\frac{d^2 f(x)}{dx^2}$. As its name implies, the second derivative is found by differentiating $\frac{df(x)}{dx}$ again. In symbols, this is expressed as

$$\frac{d^2 f}{dx^2} = \frac{d}{dx}\left(\frac{df}{dx}\right).$$

where $f(x)$ has been simplified to f for clarity.

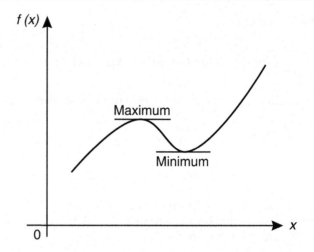

FIGURE 4.6   Graph of a function $f(x)$ that has both a maximum and a minimum.

The relationship between $\dfrac{d^2 f}{dx^2}$ and the type of stationary point is as follows:

Value of $\dfrac{d^2 f}{dx^2}$	Type of stationary point
Negative	Maximum
Zero	Point of inflexion
Positive	Minimum

A point of inflexion, as in the middle of the letter S, is both a maximum and minimum stationary point.

If

$$f(x) = x^2 - 3x$$

then

$$\frac{df(x)}{dx} = 2x - 3$$

using the rules introduced in Section 3.5.1.

It follows that $\dfrac{df(x)}{dx} = 0$ when $2x - 3 = 0$

so that $2x = 3$ or $x = \dfrac{3}{2}$.

Returning to our expression for $\dfrac{df(x)}{dx}$ we have

$$\frac{d^2 f(x)}{dx} = \frac{d}{dx}\left(\frac{df(x)}{dx}\right) = \frac{d}{dx}(2x-3) = 2$$

again using the rule from Section 3.5.1.

Since this value is positive it indicates that the stationary point at $x=\frac{3}{2}$ is a minimum.

It is worth noting that you can often see whether the value of the second derivative is positive or negative without having to evaluate it fully.

Worked Example 4.10

The total volume V of a solution of magnesium sulfate is given in terms of its molality m as

$$\frac{V}{cm^3} = 1\,001.38 - 4.86\left(\frac{m}{m^{\ominus}}\right) + 34.69\left(\frac{m}{m^{\ominus}}\right)^2$$

where $m^{\ominus}$ is the standard molality of 1 mol kg⁻¹. For what value of the molality will the volume be at a minimum?

CHEMICAL BACKGROUND

Logically, we might expect the volume to be at a minimum when no magnesium sulfate is present, and then to increase steadily as more salt is added until the solution becomes saturated. The partial molar volume of magnesium sulfate extrapolated to zero concentration has a negative value. The hydration of the Mg^{2+} and SO_4^{2-} ions causes the structure of water to break down and collapse.

Solution to Worked Example 4.10

To obtain the value of m when V will be a minimum, we need to solve the equation

$$\frac{dV}{dm} = 0$$

To differentiate the expression for V, we need to recall the standard rules from Section 3.5.1 on page 71.

$$\frac{d}{dx}(x) = 1$$

$$\frac{d}{dx}(x^2) = 2x$$

and also that the derivative of a constant (in this case 1 001.38) is zero.

The derivative is then given by

$$\frac{dV}{dm} = 0 - (4.86 \times 1) + (34.69 \times 2m)$$

$$= -4.86 + 69.38m$$

Setting $\frac{dV}{dm}$ to equal zero will allow us to find the value of m at the stationary point:

$$-4.86 + 69.38m = 0$$

To solve this equation we add 4.86 to both sides to give

$$69.38m = 4.86$$

and then divide both sides by 69.38:

$$m = \frac{4.86}{69.38} = 0.070 \, \text{mol} \, \text{kg}^{-1}$$

It would be easy to think that this was the final answer to the question; however, we have not yet shown whether this value gives a maximum or a minimum value of V. Remember that we can do this by looking at the value of $\frac{d^2V}{dm^2}$ which we obtain by differentiating our expression for $\frac{dV}{dm}$ once more:

$$\frac{d^2V}{dm^2} = \frac{d}{dm}\left(\frac{dV}{dm}\right)$$

$$= \frac{d}{dm}(-4.86 + 69.38m)$$

$$= 0 + (69.38 \times 1)$$

$$= 69.38$$

Again, no substitution of the value of m is required. Since 69.38 is a positive value, $m = 0.070 \, \text{mol} \, \text{kg}^{-1}$ gives V its minimum value.

EXERCISES

1. Evaluate:
   a. 0.46×0.921 (b) 1.25×0.65 (c) 0.362×0.6 (d) 0.402×3

2. Evaluate:
   a. $\ln 0.218$ (b) $\ln 3.74$ (c) $\log 3.19$ (d) $\log 12.17$

3. If p is defined as
   $$p = 3 - 2 \ln x$$
   where $x = \frac{y}{z}$, evaluate p when

a. $y=7.5$ and $z=2.8$ (b) $y=4.2$ and $z=3.1$
c. $y=1.8$ and $z=1.8$ (d) $y=0.2$ and $z=0.3$

4. Solve the following equations for x:
   a. $\log_x 27=3$ (b) $\log_x 64=2$ (c) $\log_x 9=2$ (d) $\log_x 8=3$
   for x.

5. Solve the following equations for x:
   a. $\log_2 x=4$ (b) $\log_3 x=2$ (c) $\log_5 x=1.5$ (d) $\log_4 x=2.5$

6. If the following quantities are constant, how are x and y related?
   a. xy (b) x^2y (c) x/y (d) x^2/y

7. The variables x and y take the following values

x	1	2	3	4	5
y	3	12	27	48	75

   Deduce the relationship between x and y.

8. If r is
   a. directly proportional to both x and y and inversely propor-
      tional to z^2
   b. directly proportional to both x and z^2 and inversely propor-
      tional to y
   c. directly proportional to both y and z^2 and inversely propor-
      tional to x
   d. directly proportional to y and inversely proportional to both
      x and z^2
      obtain a general equation which relates r to x, y and z in each
      case.

9. For the following equations, state which plot would give a straight
   line, and what the resulting values of the gradient and intercept
   would be.
   a. $y=4+6\ln x$
   b. $\log y = 2\sqrt{x}$
   c. $\ln y=2\log x-3$
   d. $\dfrac{1}{y}=x^2-7$

10. For the following equations, state which plot would give a straight
    line, and what the resulting values of the gradient and intercept
    would be.
    a. $t=s^2-5$
    b. $3w=9v+27$
    c. $a-b=14$
    d. $c^2+d=64$

11. For the following equations, state which plot would give a straight
    line, and what the resulting values of the gradient and intercept
    would be.
    a. $y=\dfrac{4}{x}+7$
    b. $\log 2y=\dfrac{1}{x^2}+6$

c. $\dfrac{1}{y^2} = \dfrac{1}{x} + \dfrac{1}{2}$

d. $y^2 = \dfrac{2}{x^2} - \dfrac{3}{4}$

12. Two variables x and y take the following values:

x	−1.5	−1.0	−0.5	0.0	0.5	1.0	1.5	2.0
y	−3.1	−2.2	−1.3	−0.4	0.5	1.4	2.3	3.2

By plotting this data, determine the relationship between x and y.

13. Two variables x and y are related by an equation of the form:

x	−3	−2	−1	0	1	2	3
y	−117.2	−37.4	−8.0	−3.8	0.4	29.8	109.6

$$y = ax^3 + b$$

Use the data below to determine the values of a and b:

14. If $f(x) = 3x^2 + 2x + 8 + \dfrac{1}{x} - \dfrac{3}{x^2}$ what is the value of:

a. $f(1)$ (b) $f(-1)$ (c) $f(-2)$ (d) $f(0.864)$?

15. At what values of x do the following functions of x have discontinuities?

a. $f(x) = \dfrac{x}{x^2 - 9}$

b. $f(x) = \dfrac{3x + 1}{2x + 3}$

c. $g(x) = \dfrac{1}{x^2} + 2$

d. $h(x) = \dfrac{1}{x - \dfrac{1}{x}}$

16. At what values of x and/or y do the following functions of x and y have discontinuities?

a. $f(x, y) = \dfrac{x + y + 1}{x^2 + y^2}$

b. $f(x, y) = \dfrac{2y}{x(x - 1)}$

c. $g(x, y) = \dfrac{x^2 - 1}{y^2 - 1}$

d. $h(x, y) = \dfrac{x^2 - y^2}{y - 4}$

17. The functions $f(x, y)$, $g(x, y)$, $h(x, y)$ and $F(x, y)$ are defined as

$$f(x,y) = \frac{x(y-1)}{y(x-1)} \quad g(x,y) = \frac{x(y^2-1)}{y(x^2-1)}$$

$$h(x,y) = \frac{x^2(y-1)}{y^2(x-1)} \quad F(x,y) = \frac{x^2(y^2-1)}{y^2(x^2-1)}$$

Each has discontinuities along two lines. Give the value of x which defines one line, and the value of y which defines the other line for (a) $f(x, y)$, (b) $g(x, y)$, (c) $h(x, y)$ and (d) $F(x, y)$.

18. For each of the functions defined by

a. $f(x)=9x^5+3x^3-2x+1$ (b) $f(x)=8x^4-2x^2+3$

c. $g(y)=2y^2-\dfrac{3}{y}$ (d) $g(y)=\dfrac{1}{y}-\dfrac{2}{y^2}+\dfrac{3}{y^3}$

determine $\dfrac{df(x)}{dx}$ and $\dfrac{d^2 f(x)}{dx^2}$ or $\dfrac{dg(y)}{dy}$ and $\dfrac{d^2 g(y)}{dy^2}$ as appropriate.

19. Locate the stationary points of the functions

a. $f(x)=2x^3-24x+7$
b. $g(x)=3x^2-2x+1$
c. $h(x)=x^2-\sqrt{x}$
d. $H(x)=x^3-3x^2+3x-1$

20. Locate and identify the stationary points of the following functions:

a. $f(x)=3x^2-7x+6$
b. $g(x)=4x^2+3x+2$
c. $G(x)=3x^3+7$
d. $H(x)=-4x^2+2x-3$

PROBLEMS

1. a.   What amount of sodium hydroxide NaOH is present in $250\,cm^3$ of a $0.25\,mol\;dm^{-3}$ solution?
   b.   What volume of this solution contains $1.00\,mol$ of sodium hydroxide?

2. What mass of sodium carbonate Na_2CO_3 is required to produce $1\,dm^3$ of a $0.30\,mol\;dm^{-3}$ solution?

3. What volume of concentrated nitric acid at $16\,mol\;dm^{-3}$ is required to produce $1\,dm^3$ of a solution of concentration $0.15\,mol\;dm^{-3}$?

4. A solution of potassium chloride in water has a concentration of $0.272\,mol\;dm^{-3}$ and a density of $1.012\,g\;cm^{-3}$. Calculate the molality of the solution.

5. a. For a solution of HCl calculate the activity of a solution of concentration 0.10 mol dm^{-3}, for which the activity coefficient is 0.796.

   b. What is the activity coefficient in a solution of HCl having a concentration 0.2 mol dm^{-3} and activity 0.15?

6. Air contains 21% oxygen. What is its partial pressure at the atmospheric pressure of 760 mm of mercury?

7. Use the Debye-Hückel equation to calculate log $\gamma_{\pm}$ for a solution of magnesium chloride of 0.03 mol dm^{-3} which has ionic strength of 0.09 mol dm^{-3}.

8. For each of the following equations, state which quantities you would plot to give a straight line, and the values of the resulting gradient and intercept:

   a. $\pi = cRT$

   Here π is the osmotic pressure which varies according to concentration c, R is the gas constant and T is the specified absolute temperature.

   b. $\log k = \log k_o + 1.02 z_A z_B \sqrt{I}$

   Here k is the rate constant for a reaction which varies according to the ionic strength I. The rate constant for the limit of unit activity coefficients is k_o and z_A and z_B are the charges on ions A and B respectively.

   c. $\Lambda = \Lambda_o - (P + Q\Lambda_o)\sqrt{c}$

   Here Λ is the molar conductivity of an electrolyte solution of concentration c, Λ_o is the molar conductivity at infinite dilution and P and Q are constants.

9. The Nernst equation gives the electromotive force E in terms of its value $E^\ominus$ under standard conditions, the reaction quotient Q, the absolute temperature T and the numerical amount of charge transferred, z, as

$$E = E^\ominus - \left(\frac{RT}{zF}\right)\ln Q$$

where R is the gas constant and F is the Faraday constant.

   In a series of measurements made on an electrochemical cell, the cell potential E was determined as a function of (a) reaction quotient Q and (b) temperature T. What information can be deduced from an appropriate straight line graph in each case?

10. Use the equation for Ostwald's dilution law:

$$K = \frac{c\left(\dfrac{\Lambda}{\Lambda_o}\right)^2}{1 - \left(\dfrac{\Lambda}{\Lambda_o}\right)}$$

to show that a graph of $1/\Lambda$ against $c\Lambda$ will be a straight line. What will be the values of the resulting gradient and intercept?

11. Fick's first law relates the diffusive flux J to the concentration gradient by the equation

$$J = -DA\frac{\partial c}{\partial x}$$

where D is the diffusion coefficient, A the area passed by the solute, c the concentration, and x the distance.

The net rate of increase of concentration with time is given by the expression

$$\frac{\partial c}{\partial t} = -\frac{1}{A}\frac{\partial J}{\partial x}$$

Use Fick's first law to eliminate J from this equation, assuming that D is independent of x.

12. a. The molar mass M of a solute is given in terms of the sedimentation coefficient s and the diffusion coefficient D:

$$M = \frac{RTs}{D(1-V_1\rho)}$$

where R is the gas constant, T the absolute temperature and V_1 the volume per unit mass of the solute. When will it *not* be possible to calculate M using this equation?

b. For a solute which undergoes a dimerization reaction with equilibrium constant K, it can be shown that

$$K = \frac{K_b(K_bm - \Delta T_b)}{(2\Delta T_b - K_bm)^2}$$

where ΔT_b is the boiling point elevation, K_b the ebullioscopic constant and m the molality. For what value of ΔT_b is K undefined according to this equation?

c. For a single 1: 1 electrolyte, the dissociation constant K is given by the expression

$$K = \frac{c^2\alpha^2}{1-\alpha}$$

where c is the concentration and α is the degree of dissociation. Under what circumstances is K not defined?

13. The total volume V of a sodium chloride solution at 25°C is given by the expression

$$\frac{V}{\text{cm}^3} = 1\,003.0 + 16.4\left(\frac{n}{\text{mol}}\right) + 2.1\left(\frac{n}{\text{mol}}\right)^{\frac{3}{2}} + 0.003\left(\frac{n}{\text{mol}}\right)^{\frac{5}{2}}$$

where n is the amount of salt in 1 kg of water. Calculate the partial molar volume V_{NaCl}, which is equal to the derivative $\dfrac{dV}{dn}$ when $n = 0.15 \, mol \, kg^{-1}$.

14. The volume V of a solution of an alcohol in water is given by the equation

$$\frac{V}{cm^3} = 1\,004.08 + 3.906\,2\left(\frac{m}{m^\ominus}\right) - 2.436\,2\left(\frac{m}{m^\ominus}\right)^2$$

where m is the molality and $m^\ominus = 1 \, mol \, kg^{-1}$. Determine the maximum volume of this solution.

15. The density $\rho(x)$ of a solution of an alcohol in water is described by the function

$$\frac{\rho(x)}{g\,cm^{-3}} = 0.982 - 0.265x + 0.306x^2$$

at a temperature of 25°C, with x being the mole fraction of alcohol in the water. Determine the value of the mole fraction for which the solution density is a maximum or a minimum. Identify the nature of the stationary point.

Kinetics

5.1 INTRODUCTION

In contrast to the thermodynamics discussed in Chapter 3, the range of mathematical techniques required for a study of kinetics is somewhat smaller. We will make much use of a technique called integration, which is essentially the reverse of the differentiation process we have already met. Analysing the rates of reactions normally involves taking a derivative, which describes the rate of change of concentration with time, and then reversing the differentiation process to obtain the original function by integrating.

5.2 USING A RATE EQUATION

For a chemical reaction between molecules A and B, the rate v of the reaction will be given by an equation of the form

$$v = k[A]^x[B]^y$$

where k is a constant known as the rate constant, which varies with temperature. [A] and [B] are the concentrations of A and B respectively, raised to respective powers x and y which are often, but not always, integers. The reaction is said to have order x with respect to A, order y with respect to B, and overall order $x+y$.

Rate always has units of mol dm^{-3} s^{-1}; since x and y take different values for different reactions it follows that k will also have different units for different reactions. Remember that units always need to be the same on either side of an equation.

Worked Example 5.1

For the reaction

$$2\,NO_{(g)} + Cl_{2(g)} \rightarrow 2\,NOCl_{(g)}$$

the rate v is given by the equation

$$v = k[NO]^2[Cl_2]$$

DOI: 10.1201/9781003043218-5

where k is the rate constant whose value is 3.84×10^{-4} dm^6 mol^{-2}s^{-1}.

    a. Calculate the rate when a vessel of volume 2.50 dm^3 contains 4.00 mol of NO and 3.50 mol Cl$_2$.

    b. What is the effect on the rate if the amount of NO is halved?

CHEMICAL BACKGROUND

Nitrosyl chloride, NOCl, can be found as an orange gas and is both toxic and corrosive. Its oxidizing properties lead to it being used as a bleaching agent in flour, and it can also be found in detergents.

Solution to Worked Example 5.1

a. We can calculate the concentrations of NO and Cl$_2$ by dividing the amount present by the volume:

$$[NO] = \frac{4.00 \text{ mol}}{2.50 \text{ dm}^3} = 1.60 \text{ mol dm}^{-3} \quad [Cl_2] = \frac{3.50 \text{ mol}}{2.50 \text{ dm}^3} = 1.40 \text{ mol dm}^{-3}$$

We can now substitute directly into the equation given

$$v = (3.84 \times 10^{-4} \text{ dm}^6 \text{mol}^{-2}\text{s}^{-1}) \times (1.60 \text{ mol dm}^{-3})^2 \times (1.40 \text{ mol dm}^{-3})$$

$$= (3.84 \times 10^{-4} \text{ dm}^6 \text{ mol}^{-2}\text{s}^{-1}) \times (2.56 \text{ mol}^2 \text{ dm}^{-6}) \times (1.40 \text{ mol dm}^{-3})$$

$$= 1.38 \times 10^{-3} \text{ mol}^{(-2+2+1)} \text{ dm}^{(6-6-3)} \text{ s}^{(-1)}$$

$$= 1.38 \times 10^{-3} \text{ mol dm}^{-3}\text{s}^{-1}$$

since, as the units of each quantity are multiplied, we add the indices each time as explained in Section 1.6.1 on page 17.

b. Since the [NO] term in the equation is squared, if we halve this the rate will change by a factor of $(\frac{1}{2})^2$ or $\frac{1}{4}$; in other words, it is reduced by a factor of 4.

$$v = \frac{1.38 \times 10^{-3} \text{ mol dm}^{-3} \text{ s}^{-1}}{4}$$

$$= 3.45 \times 10^{-4} \text{ mol dm}^{-3} \text{ s}^{-1}$$

5.3 RATES OF CHANGE

We saw in Section 3.5.1 on page 71 that we can consider both average and instantaneous rates of change. For chemical reactions, we normally monitor the change in concentration, or a quantity which can be related to the concentration, of the reactants or products with time. If we know the initial concentration, and its value after a specified time interval, it is possible to calculate the average rate of reaction.

Worked Example 5.2

In the reaction between propionaldehyde and hydrocyanic acid, the concentration of the acid was $0.090\,2$ mol dm^{-3} after 5.2 min and $0.165\,2$ mol dm^{-3} after 20.0 min. What is the average rate of reaction?

CHEMICAL BACKGROUND

Full details of this study are given in the *Journal of the American Chemical Society*, **75**, 3106, 1953. The effect of variations in the ionic strength, pH and buffer solution on the reaction rate was monitored in acetate buffers. The reaction was found to be second order.

Propionaldehyde is prepared from ethane, hydrogen and carbon monoxide by means of a homogeneously catalysed hydroformylation reaction.

Solution to Worked Example 5.2

We need to divide the change in concentration by the change in time, remembering that the change is calculated as the final value minus the initial value for both quantities

$$\text{change in concentration} = (0.165\,2 - 0.090\,2) \text{ mol dm}^{-3}$$

$$= 0.075\,0 \text{ mol dm}^{-3}$$

$$\text{change in time} = (20.0 - 5.2) \text{ min}$$

$$= 14.8 \text{ min}$$

So,

$$\text{average rate of reaction} = \frac{0.075\,0 \text{ mol dm}^{-3}}{14.8 \text{ min}}$$

$$= 5.07 \times 10^{-3} \text{ mol dm}^{-3} \text{min}^{-1}$$

This is a perfectly correct way of presenting the answer. However, it is often useful to express such quantities in terms of SI units. In this case, the SI unit of time is seconds (s), and $60\,\text{s} = 1\,\text{min}$. Substituting this information into our answer gives

$$\text{average rate of reaction} = 5.07 \times 10^{-3} \text{mol dm}^{-3} (60\,\text{s})^{-1}.$$

If we apply the power of -1 to both the number and unit in the bracket we get

$$\text{average rate of reaction} = 5.07 \times 10^{-3} \text{mol dm}^{-3} \times 60^{-1} \text{s}^{-1}$$

A power of -1 is equivalent to calculating a reciprocal ('one over a quantity') so

$$60^{-1} = \frac{1}{60}$$

and

$$\text{average rate of reaction} = \frac{5.07 \times 10^{-3} \text{ mol dm}^{-3}\text{s}^{-1}}{60}$$

$$= 8.45 \times 10^{-5} \text{ mol dm}^{-3}\text{s}^{-1}$$

Worked Example 5.3

In the reaction

$$H_{2(g)} + Br_{2(g)} \rightarrow 2HBr_{(g)}$$

the initial bromine concentration of 0.321 mol dm^{-3} falls to a value of 0.319 mol dm^{-3} after 0.005 s. Assuming that this time interval is small compared to the time taken for the reaction to reach completion, determine expressions for the rate of change of concentration of each chemical species with respect to time.

CHEMICAL BACKGROUND

Once a complex reaction such as this is underway, it is possible for subsidiary reactions to occur between the products and reactants. Because of this, the system may only be completely defined in terms of the species present and their states at the start of the reaction, and measurements of rate will then be more reliable than later on.

It is then necessary to be able to make measurements on a time scale which is short relative to that of the overall reaction.

Solution to Worked Example 5.3

We saw in Section 3.5.1 on page 71 that the value of $\frac{df(x)}{dx}$ could be considered to be the gradient of the line joining two points which had very similar values of x. Since we are told that the time interval of 0.005 s is very short in this case, the rate of change of bromine can be taken to be its instantaneous value and can therefore be represented by the derivative

$$\frac{d[Br_2]}{dt}$$

The square brackets are used to denote concentration and the subscript denoting the state is omitted for clarity.

To obtain the rate of change, we need to divide the increase in concentration by the increase in time, remembering that we always subtract initial values from final.

$$\text{change in concentration} = (0.319 - 0.321) \text{ mol dm}^{-3}$$

$$= -0.002 \text{ mol dm}^{-3}$$

Since we have defined the change in concentration as an increase, the negative value indicates that the concentration actually decreases, as we can see in the question. Since the change in time is 0.005 s the instantaneous rate of change is given by the expression

$$\frac{d[Br_2]}{dt} = \frac{-0.002 \text{ mol dm}^{-3}}{0.005 \text{ s}}$$

$$= -0.4 \text{ mol dm}^{-3}\text{s}^{-1}$$

We also need to obtain expressions for the rates of change of H_2 and HBr. Since hydrogen and bromine react in a 1:1 ratio, they are being consumed at the same rate, so we can write

$$\frac{d[H_2]}{dt} = -0.4 \text{ mol dm}^{-3} \text{ s}^{-1}$$

The equation also shows us that for every molecule of bromine consumed, two molecules of hydrogen bromide are produced. Therefore, while the concentration of bromine decreases that of hydrogen bromide increases. Since $\frac{d[Br_2]}{dt}$ has a negative value it follows that $\frac{d[HBr]}{dt}$ will have a positive value which is twice the magnitude:

$$\frac{d[HBr]}{dt} = -2\frac{d[Br_2]}{dt}$$

$$= -2 \times (-0.4 \text{ mol dm}^{-3}\text{s}^{-1})$$

$$= 0.8 \text{ mol dm}^{-3}\text{s}^{-1}$$

5.4 ZERO-ORDER REACTIONS

From our definition in Section 5.2, a zero-order reaction is one for which the sum of the powers in the rate equation is zero. Taking our general expression

$$v = k[A]^x[B]^y$$

we need to have $x + y = 0$ for a zero-order reaction. One way in which this can be obtained is when $x = y = 0$. Then

$$v = k[A]^0[B]^0$$

and since, as we saw in Section 1.6 on page 16, any number raised to a zero power is one, the expression for the rate is therefore

$$v = k \times 1 \times 1 = k$$

i.e. the rate remains constant throughout the reaction and does not depend on the concentrations of the reactants.

5.4.1 INTEGRATION

Integration is essentially the reverse of differentiation, so if we take a function, differentiate it and then integrate it, we should get back to our starting function.

Let us consider what happens if we differentiate the function $f(x) = x^2$. Remember from Section 3.5.1 on page 71 that when we differentiate we multiply by the power and reduce the power by one, so that we obtain the derivative

$$\frac{df(x)}{dx} = 2 \times x^{2-1} = 2x$$

In words, the derivative of x^2 is $2x$. Since integration is the reverse of this, we can also state that:

the integral of $2x$ is x^2

Since the 2 in this expression is simply a constant, we can divide our statement by two so that it becomes:

the integral of x is $\dfrac{x^2}{2}$

Such statements in words are cumbersome, particularly when we have more complicated expressions, and so in mathematical terminology, we write

$$\int x \, dx = \frac{x^2}{2}$$

The precise words to express this are: 'the integral of x with respect to x is equal to x squared divided by 2'. In this expression, 'dx' means 'with respect to x'.

Using similar arguments, and remembering from Section 3.5.1 on page 71 that

$$\frac{d(x^3)}{dx} = 3x^2$$

we obtain the result

$$\int x^2 dx = \frac{x^3}{3}$$

You should begin to see a pattern emerging. To integrate any polynomial type function, we simply increase the power by one and divide by the new power. This works for positive and negative powers, but not for x^{-1} (alternatively written as $\frac{1}{x}$) since this would involve dividing by zero. In mathematical terminology

$$\int x^n = \frac{x^{n+1}}{n+1} \text{ for } n \neq -1$$

There is a complication in considering integration to be the reverse process of differentiation. Applying the standard methods of differentiation gives us the derivatives

$$\frac{d}{dx}(x^3 + 1) = 3x^2$$

$$\frac{d}{dx}(x^3 + 2) = 3x^2$$

since the derivative of any constant term (1 or 2 in these examples) is zero. If we now try to reverse the process by integrating we obtain

$$\int 3x^2 = \frac{3x^3}{3} = x^3$$

and we obtain no information about the constant term. It is usual when calculating such **indefinite integrals** to include a constant of integration which is usually given the symbol C. We then obtain

$$\int 3x^2 dx = x^3 + \text{Constant}$$

or

$$\int x^2 dx = \frac{x^3}{3} + C$$

where C absorbs the divisor of 3 in the constant. C can often be calculated if we know the value of the function $f(x)$ at a certain value of x. These known values are substituted into the equation which can then be solved for C.

More commonly in chemistry, we evaluate what is called the **definite integral.** We are frequently interested in a range of values which we include as limits, and these avoid the need to calculate a constant of integration. For example, if we were asked to calculate the integral of x^3 between $x = 1$ and $x = 2$, we could express this as

$$\int_1^2 x^3 dx$$

where the lower and upper limits are included on the integration sign. To integrate, we raise the power of the function x^3 by one and divide by the new power, to obtain $\dfrac{x^4}{4}$. This expression is enclosed in square brackets with the limits outside and so we write

$$\int_1^2 x^3 dx = \left[\frac{x^4}{4} + C \right]_1^2$$

where C is the constant of integration. We then evaluate the expression at the upper limit and subtract from it the value at the lower limit, and obtain

$$\left[\frac{x^4}{4} + C \right]_1^2 = \left(\frac{2^4}{4} + C \right) - \left(\frac{1^4}{4} + C \right)$$

$$= \left(\frac{16}{4} \right) - \left(\frac{1}{4} \right)$$

$$= \frac{15}{4}$$

Note that square brackets are replaced by rounded brackets once the limits of integration have been applied. Also notice that the constants of integration have been cancelled, and we obtain an absolute value for the integral.

Worked Example 5.4

The rate equation for a zero-order reaction can be expressed as

$$-\frac{dc}{dt} = k$$

where $\dfrac{dc}{dt}$ is the rate of change of concentration c of a reactant with time t, and k is the rate constant. Obtain an expression for the concentration c at a time t assuming that the initial concentration value is c_o.

CHEMICAL BACKGROUND

A number of heterogeneous reactions obey zero-order kinetics. These include the decompositions of phosphine or ammonia on hot tungsten, shown in Figure 5.1, and the decomposition of hydrogen iodide on a hot gold wire. This is the case as long as the gas pressure is not too low.

Solution to Worked Example 5.4

As the expression given contains a derivative, we need to reverse the differentiation process by integration. The first step is to separate the variables. This is necessary before we can integrate and involves

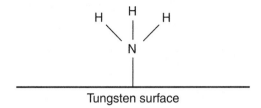

FIGURE 5.1   Ammonia decomposing on the surface of hot tungsten.

rearranging the expression so that all terms in the variable c are on one side of the equation and all terms in the variable t are on the other side. In this example, this can be done by multiplying both sides by the term dt to give

$$-dc = k\, dt$$

It is also usual to make the subject (i.e. the left-hand side of the equation) positive, and this can be achieved by multiplying both sides of the equation by -1 to give

$$dc = -k\, dt$$

We have now separated the variables and so this expression is now in a form which can be integrated. We write

$$\int dc = \int -k\, dt$$

using the integral signs to show integration.

Note that since $-k$ is a constant it can be brought outside the integral sign:

$$\int dc = -k \int dt$$

We now need to assign limits to avoid having to deal with constants of integration. Since the initial concentration is c_o, this will occur when t is zero and so our lower limits are $c = c_o$ and $t = 0$. As we require an expression for c after time t, these are set as our upper limits. This may seem strange at first, but this is a useful tool for obtaining expressions of general use. Including these limits gives us the expression

$$\int_{c_o}^{c} dc = -k \int_{0}^{t} dt$$

It may not be obvious at this stage what quantities we are going to integrate. The terms dc and dt are part of the notation of integration which say 'with respect to c' and 'with respect to t' respectively. We can multiply each of these by 1 to give

$$\int_{c_o}^{c} 1 \times dc = -k \int_{0}^{t} 1 \times dt$$

and since $c^0 = 1$ and $t^0 = 1$ (Section 1.6)

$$\int_{c_o}^{c} c^0 \, dc = -k \int_{0}^{t} t^0 \, dt$$

In this expression be careful to distinguish between c_o and c^0. We can now apply the rule that to integrate we increase the power by one and divide by the new power. Thus c^0 integrates to c and t^0 integrates to t. You may find it worthwhile remembering the rule:

$$\int dx = x + C$$

This type of standard integral should always include the constant of integration for mathematical correctness, even if we do not show it when calculating definite integrals. We now need to evaluate our integrated expression at the specified limits

$$\left[c\right]_{c_o}^{c} = -k\left[t\right]_{0}^{t}$$

If we substitute the limits for the quantities in square brackets and subtract the lower from the upper we obtain

$$c - c_o = -k(t - 0)$$

$$c - c_o = -kt$$

Alternatively, this could be multiplied throughout by –1 to give

$$c_o - c = kt$$

Worked Example 5.5

What are the units of the rate constant for a zero-order reaction?

CHEMICAL BACKGROUND

The rate constant has different units for reactions of different order. In contrast, the actual rate of a reaction will always have the units mol dm^{-3}s^{-1} since it is defined as the rate of change of a concentration with time.

Solution to Worked Example 5.5

We need to rearrange the equation

$$c_o - c = kt$$

to give an expression for k. If we divide both sides of the equation by t, we obtain

$$\frac{c_o - c}{t} = k$$

The units of k are therefore those of

$$\frac{\text{concentration}}{\text{time}}$$

which are

$$\frac{\text{mol dm}^{-3}}{\text{s}}$$

Since division is the same as multiplying by a reciprocal

$$\frac{1}{\text{s}} = \text{s}^{-1}$$

and so the SI units of the zero-order rate constant are mol dm^{-3}s^{-1}. This is as expected, since for a zero-order reaction the rate of reaction is simply equal to the rate constant.

Worked Example 5.6

The half-life of a reaction $t_{\frac{1}{2}}$ is the time taken for the concentration of reactant to fall to half its initial value. What is the half-life for a zero-order reaction?

CHEMICAL BACKGROUND

Although the definition of a half-life is given in terms of the concentration relative to its initial value, it does not actually matter when the starting concentration is taken. The time interval for a particular concentration to be halved will remain constant regardless of the starting time.

Solution to Worked Example 5.6

In this problem, we need to rearrange the integrated rate equation

$$c_o - c = kt$$

to give an expression for t. This is done by dividing both sides by k to give

$$t = \frac{c_o - c}{k}$$

From the information given in the question, when t is equal to $t_{\frac{1}{2}}$, c will be equal to $\frac{1}{2}c_0$, i.e. half the initial value. Substituting these values gives us

$$t_{\frac{1}{2}} = \frac{c_0 - \frac{1}{2}c_0}{k}$$

$$= \frac{\frac{1}{2}c_0}{k}$$

$$= \frac{c_0}{2k}$$

Worked Example 5.7

If the concentration of a reactant in a zeroth order reaction was monitored as a function of time, what graph would you plot to obtain the value of the rate constant?

CHEMICAL BACKGROUND

Some of the practicalities of monitoring such reactions are illustrated by the catalytic decomposition of hydrogen iodide on the surface of gold, which was reported by Hinshelwood and Prichard in the *Journal of the American Chemical Society* in 1925 (**127**, 1552). There is difficulty in following the pressure change during this reaction as the hydrogen iodide attacks the mercury in a manometer. This was overcome by using hydrogen gas as a buffer in the manometer since it has no effect on the rate of reaction.

Solution to Worked Example 5.7

Whenever we are analysing data, we try to plot graphs which give us straight lines. The way in which we can compare an expression with the general straight line equation $y = mx + constant$ was outlined in Section 3.6.2 on page 83. Here, x and y are the variables, and m and $constant$ are both constants representing the slope of the graph and the intercept on the y axis respectively. In this case, we need to rearrange the equation obtained in Worked Example 5.4 into this form. Note that c and t are the variables and c_0 and k are constants. If we start with

$$c - c_0 = kt$$

and add c_0 to both sides we get

$$c = -kt + c_0$$

This is now in the required form $y = mx + constant$. If we plot c on the y-axis and t on the x-axis we will obtain a straight line with gradient $-k$ and intercept c_0, as shown in Figure 5.2.

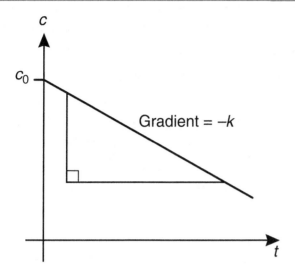

FIGURE 5.2   Graph of the integrated rate equation for a zero-order reaction.

5.5 FIRST-ORDER REACTIONS

The rate equation for a first-order reaction is of the form

$$v=k[A]$$

where [A] is the concentration of a reactant and k is the first-order rate constant.

The rate of reaction now depends on the concentration of reactant, and as this falls during the course of the reaction, so does the rate.

5.5.1 INTEGRATION OF $\dfrac{1}{x}$

We saw in Section 5.4.1 that we could integrate any expression of the form x^n as long as $n \neq -1$. The integration rule involves raising the power of x by one and dividing by the new power but, if $n=-1$, you find you need to divide by zero and the result of this is undefined. In fact, integrating x^{-1} or $\dfrac{1}{x}$ gives the expression

$$\int x^{-1}dx = \int \frac{1}{x}dx = \int \frac{dx}{x} = \ln x + C$$

containing the natural logarithm, ln x. Notice that the three forms of expressing this integral are all equivalent and you may well meet all of them.

5.5.2 RULES OF COMBINING LOGARITHMS

As noted in Section 3.6.1, in the past logarithms were used to assist in the processes of multiplication and division. If we have two numbers X and Y, then

$$\log(XY)=\log X+\log Y$$

$$\log\left(\frac{X}{Y}\right)=\log X-\log Y$$

These rules apply regardless of the base to which the logarithms are taken. They replace the process of multiplication of X by Y by the process of addition of their logs, and the process of division of X by Y by the process of subtraction of $\log Y$ from $\log X$, if the logarithms of the required numbers can be obtained.

Notice that if $X = Y$ we have

$$\log (XX)=\log X+\log X=2 \log X$$

and since

$$XX=X^2$$

It follows that

$$\log X^2=2 \log X$$

Also

$$\log\left(\frac{X}{X}\right)=\log X-\log X = 0$$

and since

$$\frac{X}{X}=1$$

it follows that

$$\log 1=0$$

Since $X^0=1$ (Section 1.6) it also follows that $\log X^0=0$.
Generally,

$$\log X^n=n \log X$$

for all values of n.

Worked Example 5.8

The rate equation for a first-order reaction can be expressed as

$$-\frac{dc}{dt}= kc$$

where $\dfrac{dc}{dt}$ is the rate of change of concentration c of a reactant with time t, and k is the rate constant. Obtain an expression for the concentration c at a time t assuming that the initial concentration value is c_o.

CHEMICAL BACKGROUND

Many reactions obey first-order kinetics. Examples include

$$2N_2O_{5(g)} \rightarrow 4NO_{2(g)} + O_{2(g)}$$

and

$$SO_2Cl_2 \rightleftharpoons SO_2 + Cl_2$$

Solution to Worked Example 5.8

Most of the stages in setting up this problem are identical to those in the case of the zero-order reaction in Worked Example 5.4. To integrate we must first separate the variables. If the terms containing variable c are to appear on the left-hand side as before it is necessary to divide both sides by c and multiply both sides by dt to give

$$-\frac{dc}{c} = k\,dt$$

before adding the integrals in symbols complete with initial and final values of c and t. This then results in the expression

$$\int_{c_o}^{c} \frac{dc}{c} = -k \int_{0}^{t} dt$$

As before,

$$\int dt = t$$

while we now know that

$$\int \frac{1}{c} dc = \ln c$$

We then obtain

$$\left[\ln c\right]_{c_o}^{c} = -k\left[t\right]_{0}^{t}$$

which on substituting the upper and lower limits into the square brackets gives

$$\ln c - \ln c_0 = -k(t-0)$$

$$= -kt$$

We saw earlier in this section that

$$\log X - \log Y = \log\left(\frac{X}{Y}\right)$$

so, in this case

$$\ln c - \ln c_o = \ln\left(\frac{c}{c_o}\right)$$

Putting this into the integrated first-order rate equation above gives us the result

$$\ln\left(\frac{c}{c_o}\right) = -kt$$

Worked Example 5.9

What are the units of the first-order rate constant?

CHEMICAL BACKGROUND

An example of a first-order rate constant is the value of $3.3\times10^{-4}s^{-1}$ for the *cis-trans* isomerization of $Cr(en)_2(OH)_2^+$. The symbol en stands for the ligand ethylenediamine, $NH_2CH_2CH_2NH_2$.

Solution to Worked Example 5.9

As in the case of Worked Example 5.5 for a zero-order reaction, we need to rearrange our integrated rate equation to make the rate constant, k, the subject. Dividing both sides of the equation

$$\ln\left(\frac{c}{c_o}\right) = -kt$$

by $-t$, gives

$$\frac{\ln\left(\frac{c}{c_o}\right)}{-t} = k$$

or

$$\frac{-\ln\left(\frac{c}{c_o}\right)}{t} = k$$

Using the rule $n \log X = \log X^n$ we have

$$\frac{\ln\left(\frac{c}{c_o}\right)^{-1}}{t} = k$$

and since from Section 1.6.5 on page 19

$$\left(\frac{c}{c_o}\right)^{-1} = \left(\frac{c_o}{c}\right)$$

we have

$$\frac{\ln\left(\dfrac{c_o}{c}\right)}{t} = k$$

Since logarithmic quantities have no units, k will have units equivalent to those of $\dfrac{1}{t}$ or t^{-1}, i.e. s^{-1}.

Worked Example 5.10

What is the half-life of a first-order reaction?

CHEMICAL BACKGROUND

The fact that different expressions emerge for the half-lives of reactions of different orders leads to a very quick way of determining whether a reaction is first order or not. The half-lives of zero- and second-order reactions depend on the initial reactant concentration, whereas the half-life of the first-order reaction does not.

Solution to Worked Example 5.10

We saw in Worked Example 5.6 that the half-life $t_{\frac{1}{2}}$ of a reaction is the time taken for the concentration to fall to half its initial value. Rearranging the integrated rate equation

$$\frac{\ln\left(\dfrac{c_o}{c}\right)}{t} = k$$

for a first-order reaction to give an expression for t gives

$$t = \frac{1}{k}\ln\left(\frac{c}{c_o}\right)$$

Substituting $t = t_{\frac{1}{2}}$ and $c = \dfrac{1}{2}c_o$ now gives

$$t_{\frac{1}{2}} = \frac{1}{k}\ln\left(\frac{c_o}{\dfrac{1}{2}c_o}\right)$$

$$= \frac{\ln 2}{k}$$

since the c_o terms cancel and $1 \div \dfrac{1}{2} = 2$.

Worked Example 5.11

A determination of the pressure of gaseous N_2O_5 as a function of time gave the values

$\dfrac{t}{s}$	0	1200	4800	7200
$\dfrac{p}{mm\,Hg}$	350	190	34	10

If the decomposition of N_2O_5 is a first-order reaction, what is the value of the rate constant?

CHEMICAL BACKGROUND

This is a frequently quoted example of a first-order reaction, the equation for the decomposition being

$$2N_2O_{5(g)} \rightarrow 4NO_{2(g)} + O_{2(g)}$$

The total pressure can be measured by connecting the flask to a manometer, but it is then necessary to calculate the contribution from N_2O_5.

Solid N_2O_5 consists of NO_2^+ and NO_3^- ions. It is colourless and stable below 0°C. At higher temperatures, it decomposes slowly into N_2O_4 and O_2.

Pressure is a useful quantity to determine since it is proportional to concentration. If we consider the ideal gas equation

$$pV = nRT$$

and divide both sides by V, we obtain

$$p = \left(\frac{n}{V}\right)RT$$

The quantity $\dfrac{n}{V}$ has the units of mol per unit volume, so is actually a measure of concentration c. We can therefore write

$$p = cRT$$

and since R is a constant at a constant temperature p will be proportional to c.

Solution to Worked Example 5.11

Since this is a first-order reaction, we will use the integrated rate equation

$$\ln c - \ln c_0 = -kt$$

If the concentration c is proportional to pressure p, we can write

$$c = Kp$$

and

$$c_o = Kp_o$$

where K (in this case equal to $\dfrac{1}{RT}$) is the constant of proportion-ality as explained in Section 3.7.4 on page 95 and p_o is the initial pressure. Substituting into the integrated rate equation then gives

$$\ln(Kp) - \ln(Kp_o) = -kt$$

We saw in Section 5.5.2 on page 161 that

$$\ln(XY) = \ln X + \ln Y$$

and so

$$\ln(Kp) = \ln K + \ln p$$

and

$$\ln(Kp_o) = \ln K + \ln p_o$$

can be substituted into the left-hand side of the equation to give

$$(\ln K + \ln p) - (\ln K + \ln p_o) = -kt$$

The brackets can be removed, ensuring that the minus sign is applied to both terms in the second bracket, to give

$$\ln K + \ln p - \ln K - \ln p_o = -kt$$

The terms $\ln K$ and $-\ln K$ cancel, to leave

$$\ln p - \ln p_o = -kt$$

Adding $\ln p_o$ to both sides of the equation gives

$$\ln p = -kt + \ln p_o$$

Since p_o is a constant $\ln p_o$ is also a constant, and so we now have an equation which is of the form $y = mx + c$, as discussed in Section 3.6.2 on page 83. Comparing corresponding quantities in the two forms of the equation shows that if $\ln p$ is plotted on the y-axis and t on the x-axis, we will obtain a straight line graph with gradient $-k$.

The mechanics of graph plotting were outlined in Section 3.6.3 on page 85. We first need to draw up a table containing values of ln p and t. We will actually calculate the values of $\ln\left(\dfrac{p}{\text{mm Hg}}\right)$ since we can only take the logarithm of a quantity which has no units. This actually has the effect of introducing a further constant, but does not affect the value of the gradient.

The first value in the table has

$$\frac{p}{\text{mm Hg}} = 350$$

so if we use a calculator we obtain

$$\ln\left(\frac{p}{\text{mm Hg}}\right) = \ln 350$$

$$= 5.86$$

We can obtain other values in a similar fashion, and the data we need to plot can be tabulated as

$\dfrac{t}{s}$	0	1200	4800	7200
$\dfrac{p}{\text{mm Hg}}$	350	190	34	10
$\ln\left(\dfrac{p}{\text{mm Hg}}\right)$	5.86	5.25	3.53	2.30

The graph of $\ln\left(\dfrac{p}{\text{mm Hg}}\right)$ against t is shown in Figure 5.3, from which we can obtain the gradient.

$$\text{gradient} = \frac{\text{increase in} \ln\left(\dfrac{p}{\text{mm Hg}}\right)}{\text{increase in} \, t}$$

$$= \frac{2.00 - 5.86}{(7900 - 0)\text{s}}$$

$$= \frac{-3.86}{7900\,\text{s}}$$

$$= -4.9 \times 10^{-4}\,\text{s}^{-1}$$

We have already seen that the gradient is equal to $-k$, so

$$k = 4.9 \times 10^{-4}\text{s}^{-1}$$

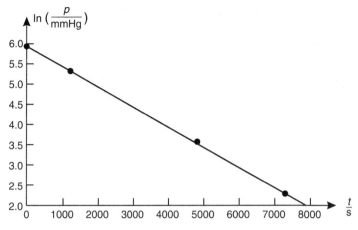

FIGURE 5.3   Plot of $\ln\left(\dfrac{p}{mm\ Hg}\right)$ against time for the decomposition of N_2O_5.

5.6 SECOND-ORDER REACTIONS

If we consider the generalized expression for the rate v of a reaction

$$v=k[A]^x[B]^y$$

there are two ways in which it is possible to achieve an overall order of 2, i.e. to satisfy the equation

$$x+y=2$$

Case 1: $x=2, y=0$
   Note that a similar argument applies if $x=0, y=2$.
   If $x=2$ and $y=0$ the rate equation is

$$v=k[A]^2$$

This would arise in a reaction of the type

$$A+B \rightarrow products$$

in which the starting concentrations of A and B were equal.
Case 2: $x=y=1$
   If $x=y=1$, the equation is

$$v=k[A][B]$$

The first of these two cases is rather simpler to deal with than is the second. To analyse second-order kinetics in the situation when the starting concentrations of the two reactants are not equal, we need some additional mathematical techniques in the form of partial fractions. These will be introduced in Section 5.6.1.

Worked Example 5.12

The rate equation for a second-order reaction in which the initial concentrations of the reactants are equal, or in which there is a single reactant, can be expressed as

$$-\frac{dc}{dt} = kc^2$$

where $\frac{dc}{dt}$ is the rate of change of concentration c of a reactant with time t, and k is the rate constant. Obtain an expression for the concentration c at a time t, assuming that the initial concentration value is c_o.

CHEMICAL BACKGROUND

An example of a reaction which obeys second-order kinetics is the dimerization of butadiene

$$2\,C_4H_{6(g)} \rightarrow C_8H_{12(g)}$$

for which

$$v = -\frac{d[C_4H_6]}{dt} = k[C_4H_6]^2$$

Solution to Worked Example 5.12

The initial stages involve rearranging the given equation to separate the terms, i.e. to have all terms in c on the left-hand side and all those in t on the right-hand side. Then we can integrate and assign limits as in Worked Examples 5.4 and 5.8. This results in us being required to solve the equation

$$\int_{c_o}^{c} \frac{dc}{c^2} = -k \int_0^t dt$$

The integration on the right-hand side is straightforward and identical to the previous problems.

$$\int_0^t dt = t$$

Integration of $\frac{1}{c^2}$ requires a little more thought, but is not difficult once we realize that $\frac{1}{c^2}$ can be written as c^{-2}. Raising the power by one gives us c^{-1} (not c^{-3} which would be a reduction of the power by one – this is a common error) so we need to divide c^{-1} by -1. Note that dividing by -1, is the same as multiplying by -1 because

$$x \div -1 = \frac{x}{-1} = -x \text{ and } x \times -1 = -x$$

Consequently, we need to evaluate the expression

$$\left[-\frac{1}{c} \right]_{c_0}^{c} = -k[t]_0^t$$

If we substitute the upper limits and then subtract the value obtained with the lower limits, we get

$$-\frac{1}{c} - \left(-\frac{1}{c_0} \right) = -k(t - 0)$$

which, if we multiply every term by -1, becomes

$$\frac{1}{c} - \frac{1}{c_0} = kt$$

Worked Example 5.13

What are the units of the second-order rate constant?

CHEMICAL BACKGROUND

An example of the second-order rate constant is the value of 2.42×10^{-2} dm³mol⁻¹s⁻¹ for the reaction

$$H_2 + I_2 \rightarrow 2HI$$

An alternative method for synthesizing hydrogen iodide is by reducing iodine with hydrogen sulfide. Hydrogen compounds generally become less stable as a group of the periodic table is descended, so HF will be more stable than HI.

Solution to Worked Example 5.13

As with other integrated rate equations, we need to rearrange them to make the rate constant k the subject of the equation. We can do this by dividing both sides of the equation by t, to give

$$k = \frac{1}{t} \left(\frac{1}{c} - \frac{1}{c_0} \right)$$

The units we would obtain from substituting into this equation would therefore be

$$\frac{1}{s} \left(\frac{1}{mol\,dm^{-3}} \right)$$

Note that

$$\frac{1}{s} = s^{-1} \text{ and } \frac{1}{dm^{-3}} = dm^3$$

Then the units can be expressed as $dm^3\,mol^{-1}\,s^{-1}$. We write the units with the positive powers before those with the negative powers because, when expressed verbally we replace negative powers by the word 'per'. Thus $dm^3\,mol^{-1}\,s^{-1}$ would be spoken as: decimeters cubed per mole per second.

Worked Example 5.14

If the concentration of a reactant in a second-order reaction, with equal concentrations of reactants, was monitored as a function of time, what graph would you plot to obtain the value of the rate constant?

CHEMICAL BACKGROUND

The analysis of kinetics in this example would also apply to a reaction with a single reactant, such as the disproportionation of $HBrO_2$ in acidic solution according to the equation

$$2\,HBrO_2 \rightarrow HOBr + HBrO_3$$

or the reaction

$$2\,I \rightarrow I_2$$

in the gas phase.

Solution to Worked Example 5.14

In order to obtain a value of the rate constant k, we need to plot the data in such a way that we obtain a straight line. This can be done by rearranging the integrated rate equation above into the form of the equation of a straight line graph, $y = mx + c$, as outlined in Section 3.6.2 on page 83.

Starting with the equation

$$\frac{1}{c} - \frac{1}{c_o} = kt$$

we can add the constant term $\dfrac{1}{c_o}$ to both sides to give

$$\frac{1}{c} = kt + \frac{1}{c_o}$$

This is now in the form we want, with $\dfrac{1}{c}$ corresponding to y and t corresponding to x. If a graph of $\dfrac{1}{c}$ against t is plotted, we will

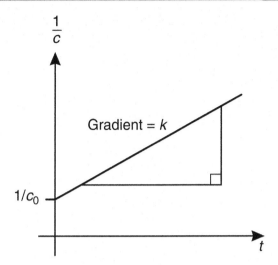

FIGURE 5.4   Graph of the integrated rate equation for a second-order reaction.

obtain a straight line with gradient k and intercept $\dfrac{1}{c_o}$ as shown in Figure 5.4.

Worked Example 5.15

During the hydrolysis of methyl acetate, aliquots of the reaction mixture were withdrawn at intervals and titrated with sodium hydroxide, to give the following results:

$\dfrac{t}{\text{min}}$	5.64	20.69	45.74	75.76
$\dfrac{V}{\text{cm}^3}$	26.52	27.98	29.88	31.99

where V is the titre of sodium hydroxide. Determine the order of this reaction.

CHEMICAL BACKGROUND

The data given refer to the acid hydrolysis of methyl acetate, which can be represented by the equation

$$CH_3COOCH_3 + H_2O \rightleftharpoons CH_3COOH + CH_3OH$$

The carbonyl oxygen becomes protonated and so the carbonyl carbon atom is more susceptible to nucleophilic attack by the water molecule. There are likely to be several tetrahedral intermediates in this reaction.

Methyl acetate is an ester which can be formed by the reaction of acetic anhydride with methanol. It melts at −98°C and boils at 57.5°C.

Solution to Worked Example 5.15

We have seen earlier that there are three distinct plots which will give a straight line in reactions we are likely to meet. To summarize, we need to plot

- c against t for a zero-order reaction
- $\ln c$ against t for a first-order reaction
- $\dfrac{1}{c}$ against t for a second-order reaction

To determine the order of a reaction such as this, we need to draw the three graphs and identify which one gives the straight line. Hopefully, this will be easy to spot.

In this example, there are further complications. We are not given the concentration of methyl acetate, but the titre V of sodium hydroxide required to neutralize the methanoic acid formed. However, since the rate of increase of methanoic acid is numerically equivalent to the rate of decrease of methyl acetate, these quantities are proportional and we can write

$$V = Kc$$

where K is the constant of proportionality and so for the initial volume V_o and initial concentration c_o

$$V_o = Kc_o$$

If we now substitute into the integrated rate equation for a zero-order reaction

$$c = -kt + c_o$$

we have

$$\frac{V}{K} = -kt + \frac{V_o}{K}$$

We can multiply both sides by the proportionality constant K to give

$$V = -kKt + V_o$$

so in fact a plot of V against t will be a straight line having gradient kK and intercept V_o.

Similarly, it is possible to show that straight line plots will be obtained for a reaction of appropriate order if a quantity which is proportional to the concentration is used instead of the concentration itself. In this problem, it is therefore possible to use the titre values directly, rather than converting them to concentrations.

A further complication is that we do not have the values of V_0, the titre value at zero time. However, as the time values are all relative, we can subtract the first value of 5.64 from each, which is equivalent to assuming that the reaction began 5.64 minutes later. If we do this, the data we need to use become

$\dfrac{t'}{\text{min}}$	5.64	20.69	45.74	75.76
$\dfrac{t}{\text{min}}$	0.00	15.05	40.10	70.12
$\dfrac{V}{\text{cm}^3}$	26.52	27.98	29.88	31.99

where we have denoted the original uncorrected time values as t'.

To consider the first- and second-order plots, we need the values of

$$\ln\left(\frac{V}{\text{cm}^3}\right) \text{ and } \frac{1}{V}$$

If we consider the first point, we have

$$V = 26.52 \text{ cm}^3$$

$$\frac{V}{\text{cm}^3} = 26.52$$

$$\ln\left(\frac{V}{\text{cm}^3}\right) = 3.278$$

$$\frac{1}{V} = 0.037\,71\,\text{cm}^{-3}$$

The full table with these quantities is as follows:

t/min	0.00	15.05	40.10	70.12
V/cm³	26.52	27.98	29.88	31.99
$\ln(V/\text{cm}^3)$	3.278	3.331	3.397	3.465
V^{-1}/cm⁻³	0.037 71	0.035 74	0.033 46	0.031 26

The graphs of V against t, $\ln V$ against t and $\dfrac{1}{V}$ against t are shown in Figure 5.5. Since only the plot of $\ln V$ against t gives a straight line, the reaction must be first order.

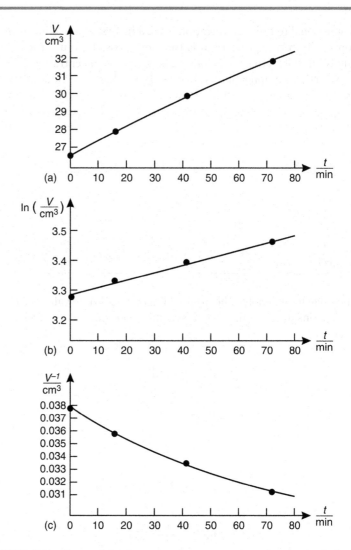

FIGURE 5.5 Plots of the integrated rate equations for the hydrolysis of methyl acetate: (a) zero-order, (b) first-order, (c) second-order.

5.6.1 PARTIAL FRACTIONS

Partial fractions are useful for expressing a fractional product as the sum of two fractions, which in some situations may be easier to deal with. For example, if we have an expression such as

$$\frac{1}{(x+1)(x+2)}$$

it may be preferable to express it as the sum of two fractions such as

$$\frac{A}{x+1}+\frac{B}{x+2}$$

where A and B are constants. Determining the partial fractions, in this case, is simply a matter of finding values for the constants A and B. Since these expressions are identical, we can write the **identity**

$$\frac{1}{(x+1)(x+2)} \equiv \frac{A}{(x+1)} + \frac{B}{(x+2)}$$

Notice the use of the $\equiv$ symbol here. It is used instead of $=$ because this expression is true for all values of x. An equation, on the other hand, is only true for certain values of x. We will exploit this difference in the following analysis.

If we now consider the right-hand side of this expression, we can multiply the top and bottom of the first term by $(x+2)$ and the top and bottom of the second term by $(x+1)$:

$$\frac{1}{(x+1)(x+2)} \equiv \frac{A(x+2)}{(x+1)(x+2)} + \frac{B(x+1)}{(x+2)(x+1)}$$

because multiplying both the top and bottom of a fraction by the same quantity does not change its value, as explained in Section 1.5.1 on page 11. The two terms on the right-hand side now have the same denominators so we can now write

$$\frac{1}{(x+1)(x+2)} \equiv \frac{A(x+2)+B(x+1)}{(x+1)(x+2)}$$

Multiplying both expressions by $(x+1)(x+2)$, gives

$$1 \equiv A(x+2)+B(x+1)$$

Since this is an identity, it will be true for any value of x. If we choose to set $x=-1$, the bracket containing $x+1$ will become zero and we have

$$A(-1+2) \equiv 1$$

$$A \equiv 1$$

Similarly if we set $x=-2$, the bracket containing $x+2$ becomes zero and we have

$$B(-2+1) \equiv 1$$

$$-B \equiv 1$$

$$B \equiv -1$$

Substituting for A and B in the general expression for the partial fractions then gives

$$\frac{1}{(x+1)(x+2)} \equiv \frac{1}{x+1} + \frac{-1}{x+2}$$

$$\equiv \frac{1}{x+1} - \frac{1}{x+2}$$

There are more complicated cases of partial fractions, but this knowledge is sufficient for the analysis of second-order reactions required here.

5.6.2 DIFFERENTIATION OF LOGARITHMIC FUNCTIONS AND INTEGRATION OF FRACTIONS

We have seen previously in Section 5.5.1 on page 161 that

$$\int \frac{dx}{x} = \ln x + C$$

where C is the constant of integration. If integration of $\frac{1}{x}$ gives $\ln x$, and differentiation is the reverse of integration (Section 5.4.1 on page 154), differentiation of $\ln x$ will give $\frac{1}{x}$. Expressed mathematically

$$\frac{d}{dx}(\ln x) = \frac{1}{x}$$

It is also true that

$$\frac{d}{dx}(\ln(ex+f)) = \frac{e}{ex+f}$$

where e and f are constants so that, for example

$$\frac{d}{dx}(\ln(3x+2)) = \frac{3}{3x+2}$$

Since integration is the reverse of differentiation, we also have

$$\int \frac{e\,dx}{ex+f} = \ln(ex+f) + C$$

and so

$$\int \frac{ex}{ex+f} = \frac{1}{e}\ln(ex+f) + C$$

where C is the constant of integration.

There is a complication in that we can only take the logarithm of a positive number, and strictly speaking the correct expression for dealing with general integrations of this type is

$$\int \frac{dx}{ex+f} = \frac{1}{e}\ln|ex+f|+C$$

The vertical bars simply denote that we take the modulus of the quantity enclosed between them. This is the value taking no account of its sign, as we saw previously in Section 4.6 on page 131. In the examples we meet in chemistry the modulus symbols are frequently omitted, so we would write

$$\int \frac{dx}{ex+f} = \frac{1}{e}\ln(ex+f)+C$$

Worked Example 5.16

The rate equation for a second-order reaction in which the initial concentrations of the reactants are different can be expressed as

$$-\frac{da}{dt} = kab$$

where $\frac{da}{dt}$ is the rate of change of concentration a of one reactant with time t, b is the concentration of the second reactant, and k is the rate constant. Obtain an expression for the concentrations a and b at a time t assuming that the initial concentration values are a_0 and b_0.

CHEMICAL BACKGROUND

An example of such a reaction is that between ethylene bromide and potassium iodide according to the reaction

$$C_2H_4Br_2 + 3\ KI \rightarrow C_2H_4 + 2\ KBr + KI_3$$

for which the rate v is given by

$$v = k\ [C_2H_4Br_2]\ [KI]$$

Solution to Worked Example 5.16

As previously, we begin by separating the variables, i.e. rearranging the expression given so that all the concentration terms are on the left, with the rate constant and the time on the right. This gives us the expression

$$\int \frac{da}{ab} = -k\int dt$$

where the rate constant k is taken outside the integral on the right-hand side. We immediately have a problem, since we need to integrate

$$\frac{1}{ab}$$

with respect to a and this cannot be done as the expression stands because it also contains the variable b. To overcome this difficulty, remember that the generalized equation for this reaction is

$$A+B \rightarrow \text{products}$$

The initial concentrations of A and B are given as a_o and b_o respectively. Suppose that after time t the concentration of A has reduced by x; if A and B react in an equimolar ratio then the concentration of B will also have reduced by x and we can write for the concentrations of A and B

$$a=a_o-x$$

and

$$b=b_o-x$$

Since a_o and b_o are constants, we have now expressed the concentrations of A and B in terms of a single common variable x. If we differentiate the first equation, since a_o is a constant then

$$\frac{da}{dx}=-1$$

and so we can replace da by $-dx$ in the integration.

We are now able to substitute for a, b and da to obtain

$$\int \frac{dx}{(a_o-x)(b_o-x)} = k \int dt$$

noticing that the minus signs on either side cancel. This expression now needs to be integrated between appropriate limits. Clearly when $t=0$ none of A or B has reacted so $x=0$, so we take these as the lower limits in the integration. We want to obtain an expression giving x after time t so we use these values as our upper limits:

$$\int_0^x \frac{dx}{(a_o-x)(b_o-x)} = k \int_0^t dt$$

Integrating the right-hand side is straightforward:

$$\int_0^t dt = [t]_0^t = t-0=t$$

However, we now have the problem of integrating the expression on the left, which is not so easy. We saw earlier how an expression of this type could be written in terms of two partial fractions. As before, we start by setting up an identity in terms of two unknown constants A and B:

$$\frac{1}{(a_o-x)(b_o-x)} \equiv \frac{A}{(a_o-x)} + \frac{B}{(b_o-x)}$$

We multiply the first term, top and bottom, by (b_o-x) and the second term, top and bottom, by (a_o-x); neither of these operations changes the value of the expression but it does produce fractions with the same denominators:

$$\frac{1}{(a_o-x)(b_o-x)} \equiv \frac{A(b_o-x)}{(a_o-x)(b_o-x)} + \frac{B(a_o-x)}{(b_o-x)(a_o-x)}$$

Since the quantity on the bottom is the same for every term in the expression we multiply all terms by $(a_o-x)(b_o-x)$ to give

$$1 \equiv A(b_o-x) + B(a_o-x)$$

This expression is true for all values of x, so we set $x = b_o$ to make the first bracket zero and to leave

$$1 \equiv B(a_o-b_o)$$

This can be rearranged to give

$$B \equiv \frac{1}{a_o-b_o}$$

Similarly, setting $x = a_o$ makes the second bracket zero and we have

$$1 \equiv A(b_o-a_o)$$

which can be rearranged to give

$$A \equiv \frac{1}{b_o-a_o}$$

It is worth noticing that $A = -B$, as this will allow some simplification later.

Substituting in our original identity for the partial fractions gives

$$\frac{1}{(a_o-x)(b_o-x)} \equiv \frac{1}{(b_o-a_o)(a_o-x)} + \frac{1}{(a_o-b_o)(b_o-x)}$$

If we multiply the top and bottom of the second term on the right by -1 this becomes

$$\frac{1}{(a_o-x)(b_o-x)} \equiv \frac{1}{(b_o-a_o)(a_o-x)} - \frac{1}{(b_o-a_o)(b_o-x)}$$

We can then introduce brackets on the right-hand side and take the common factor of $(b_0 - a_0)$ outside the new brackets to give

$$\frac{1}{(a_0 - x)(b_0 - x)} = \frac{1}{(b_0 - a_0)}\left(\frac{1}{(a_0 - x)} - \frac{1}{(b_0 - x)}\right)$$

Bearing in mind that a_0 and b_0, and consequently $b_0 - a_0$ and $\dfrac{1}{b_0 - a_0}$, are constants, we can now write our equation to be integrated as

$$\frac{1}{b_0 - a_0} \int_0^x \left(\frac{1}{(a_0 - x)} - \frac{1}{(b_0 - x)}\right) dx = k \int_0^t dt$$

or

$$\frac{1}{b_0 - a_0} \int_0^x \frac{dx}{a_0 - x} - \frac{1}{b_0 - a_0} \int_0^x \frac{dx}{b_0 - x} = k \int_0^t dt$$

We saw earlier that

$$\int \frac{dx}{ex + f} = \frac{1}{e} \ln(ex + f) + C$$

Comparing this with the first integral

$$\int \frac{dx}{a_0 - x}$$

shows that $e = -1$ and $f = a_0$. Consequently

$$\int \frac{dx}{a_0 - x} = \frac{1}{-1} \ln(-x + a_0)$$

$$= -\ln(a_0 - x) + C$$

This allows us to write the following expression for the definite integral:

$$\int_0^x \frac{dx}{a_0 - x} = \left[-\ln(a_0 - x)\right]_0^x$$

Substituting for the upper limit and then subtracting the expression with the lower limit gives

$$\int_0^x \frac{dx}{a_0 - x} = \left[-\ln(a_0 - x)\right] - \left[-\ln(a_0 - 0)\right]$$

$$= -\ln(a_0 - x) + \ln a_0$$

since the adjacent negative signs combine to give a positive one. We saw in Section 5.5.2 on page 161 that

$$\ln D - \ln E = \ln\left(\frac{D}{E}\right)$$

so that

$$\ln a_o - \ln(a_o - x) = \ln\left(\frac{a_o}{a_o - x}\right)$$

and

$$\int_0^x \frac{dx}{a_o - x} = \ln\left(\frac{a_o}{a_o - x}\right)$$

Similarly, we obtain

$$\int_0^x \frac{dx}{b_o - x} = \ln\left(\frac{b_o}{b_o - x}\right)$$

and finally we have

$$\int_0^t dt = t$$

as we have already seen several times for the other rate equations. Putting the results of these three integrations into the overall expression leads to

$$\frac{1}{b_o - a_o}\left(\ln\left(\frac{a_o}{a_o - x}\right) - \ln\left(\frac{b_o}{b_o - x}\right)\right) = kt$$

Once again we have the difference between two logarithmic terms

$$\ln D - \ln E = \ln\left(\frac{D}{E}\right) = \ln\left(DE^{-1}\right)$$

so we can write

$$\frac{1}{b_o - a_o}\left(\ln\left(\frac{a_o}{a_o - x}\right)\left(\frac{b_o}{b_o - x}\right)^{-1}\right) = kt$$

which can also be expressed more neatly as

$$\frac{1}{(b_o - a_o)}\left(\ln\frac{a_o(b_o - x)}{b_o(a_o - x)}\right) = kt$$

since

$$\left(\frac{b_o}{b_o-x}\right)^{-1}=\frac{b_o-x}{b_o}$$

The integrations above have been performed with respect to the variable x since this could be related to the concentrations of both A and B. However, it is now possible to eliminate x from the expression since we set

$$a=a_o-x$$

and

$$b=b_o-x$$

where a and b represent concentrations after time t. This gives the expression as

$$\frac{1}{(b_o-a_o)}\ln\left(\frac{a_ob}{ab_o}\right)=kt$$

Since

$$\left(\frac{a_ob}{ab_o}\right)=\left(\frac{ab_o}{a_ob}\right)^{-1}$$

this could also be written as

$$\frac{1}{(b_o-a_o)}\ln\left(\frac{ab_o}{a_ob}\right)^{-1}=kt$$

From Section 5.5.2 on page 161 we know that

$$\ln x^{-1}=-\ln x$$

and so

$$\frac{-1}{(b_o-a_o)}\ln\left(\frac{ab_o}{a_ob}\right)=kt$$

Multiplying this expression top and bottom by -1 gives the final expression

$$\frac{1}{(a_o-b_o)}\ln\left(\frac{ab_o}{a_ob}\right)=kt$$

The derivation of this expression requires several mathematical techniques and not a few "tricks of the trade" to get intermediate expressions into the required form; fortunately, few of the mathematical processes we have met or will meet are of such length or complexity! You are also unlikely to need to perform this series of calculations in any other chemical context.

5.7 THE ARRHENIUS EQUATION

So far we have considered the rate constant k to be constant under specified conditions. This is in fact true as long as the temperature remains constant but, as you might expect, the rate of a reaction increases with the temperature. The relationship between rate constant and temperature is described by an exponential function.

5.7.1 THE EXPONENTIAL FUNCTION

The exponential function can be defined in terms of our usual notation as

$$f(x) = e^x$$

where e has the value 2.718 3 to four decimal places. This value is not exact, but we rarely need to worry about the numerical value of e. When complicated expressions are involved, you will also see the exponential function written as

$$f(x) = \exp(x)$$

This has precisely the same meaning as the previous equation and is simply a useful device for improving the clarity of some mathematical expressions.

Scientific calculators allow easy calculation of exponential functions.

It is worth noting that it is only possible to take the exponential of a quantity which has no units, in other words, a pure number. The absence of units in an expression should always be verified before attempting to calculate an exponential term in a calculation.

5.7.2 INVERSE FUNCTIONS

The inverse of a function $f(x)$ is generally known as $\mathrm{arc}f(x)$ and simply reverses the effect of the original function. For example, if a function $f(x)$ is defined by

$$f(x) = x + 10$$

this means 'take a value of x and add 10 to it'. To reverse this we must therefore subtract 10 and we write

$$\mathrm{arc}\, f(x) = x - 10$$

Notice that

$$\mathrm{arc}\, f(f(x)) = f(x) - 10$$

$$= (x+10) - 10$$

$$= x$$

which is a general result for a function and its inverse.

The natural logarithmic and exponential functions are actually the inverses of each other, so that

$$\mathrm{arc}\, \ln(x) = e^x \text{ and arc } e^x = \ln x$$

It also follows that

$$e^{\ln x} = x \text{ and } \ln e^x = x$$

Similarly

$$10^{\log x} = x \text{ and } \log 10^x = x$$

In other words the inverse of $\log x$ is 10^x.

Worked Example 5.17

The value of the rate constant k for the decomposition of nitrogen dioxide was found to vary with absolute temperature T according to the Arrhenius equation

$$k = A \exp\left(\frac{-E_a}{RT}\right)$$

where A is a constant called the pre-exponential factor, E_a is the activation energy and R is the gas constant which can be taken as $8.31\, \mathrm{J\, K^{-1} mol^{-1}}$. Use the data below to determine the value of the activation energy E_a:

$\dfrac{T}{K}$	593	604	628	652	657
$\dfrac{k}{\mathrm{dm^3\, mol^{-1}\, s^{-1}}}$	0.523	0.751	1.70	4.01	5.01

CHEMICAL BACKGROUND

For some reactions involving free atoms or radicals, there is a very small activation energy and a more accurate treatment of the temperature dependence must be used. This frequently involves using an equation of the form

$$k = AT^n \exp\left(-\frac{E}{RT}\right)$$

where the value of n depends on the nature of the reaction and the theory being used in the analysis. The activation energy E in this equation is related to that in the Arrhenius equation, E_a, by the equation

$$E = E_a - nRT$$

Solution to Worked Example 5.17

A straightforward plot of k against T is shown in Figure 5.6(a) and is obviously curved. As usual, we would like to obtain a linear plot, and this can be done by taking the natural logarithm of each side of the Arrhenius equation. This gives us

$$\ln k = \ln\left(A\exp\left(-\frac{E_a}{RT}\right)\right)$$

The right-hand side of this equation now has the logarithm of a product, and since

$$\ln(XY) = \ln X + \ln Y$$

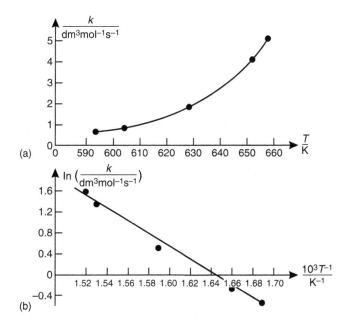

(a)

(b)

FIGURE 5.6   Plot of (a) rate constant k against temperature T, and (b) $\ln k$ against $1/T$ for the decomposition of nitrogen dioxide.

as we saw in Section 5.5.2 on page 161 we can rewrite the equation as

$$\ln k = \ln A + \ln\left(\exp\left(-\frac{E_a}{RT} \right) \right)$$

We have also seen that $\ln e^x = x$ and so our modified Arrhenius equation becomes

$$\ln k = \ln A - \frac{E_a}{RT}$$

Reversing the order of the terms on the right, and grouping the constants E_a and R in brackets gives us

$$\ln k = -\left(\frac{E_a}{R} \right)\left(\frac{1}{T} \right) + \ln A$$

which is equivalent to the general equation of a straight line $y = mx + c$ with $\ln k$ being equivalent to y, $-\frac{E_a}{R}$ being equivalent to the gradient m, $\frac{1}{T}$ being equivalent to x, and $\ln A$ being equivalent to the intercept c (Section 3.6.2 on page 83). We therefore need to plot $\ln k$ against $\frac{1}{T}$ to obtain a straight line with gradient $-\frac{E_a}{R}$.

Since we can only take the logarithm of a pure number without units we will actually calculate values of

$$\ln\left(\frac{k}{dm^3\,mol^{-1}\,s^{-1}} \right)$$

Strictly speaking, this is the quantity which should be on the left-hand side of the equation above. However, it is tedious to keep writing this so we tend to use the equation as shown and remember that we need to deal with the units in this way. For the first point,

$$k = 0.523\ dm^3\,mol^{-1}\,s^{-1}$$

so

$$\frac{k}{dm^3\,mol^{-1}\,s^{-1}} = 0.523$$

and using a calculator gives

$$\ln\left(\frac{k}{dm^3\,mol^{-1}\,s^{-1}} \right) = -0.648$$

The temperature values are given in units of K so directly give the values of T. The first point has $T = 593$ K and the calculator gives

$$T^{-1} = \frac{1}{T} = 1.69 \times 10^{-3} \, \mathrm{K}^{-1}$$

It is worth looking at how this expression can be rearranged to assist in the tabulation of data. It is neater to incorporate the power of ten into the table heading, rather than including it for every value. We then have, for example, $\dfrac{10^{3} T^{-1}}{\mathrm{K}^{-1}} = 1.69$, which rearranges to give $T^{-1} = 1.69 \times 10^{-3}$ K^{-1}. The remaining values of T^{-1} are calculated in the same way and can be tabulated as follows.

$\ln\left(\dfrac{k}{\mathrm{dm^3\,mol^{-1}\,s^{-1}}}\right)$	−0.648	−0.286	0.531	1.39	1.61
$\dfrac{10^{3} T^{-1}}{\mathrm{K}^{-1}}$	1.69	1.66	1.59	1.53	1.52

The plot of $\ln\left(\dfrac{k}{\mathrm{dm^3\,mol^{-1}\,s^{-1}}}\right)$ against T^{-1} is shown in Figure 5.6(b). From this, we can calculate

$$\mathrm{gradient} = \frac{\text{increase in } \ln k}{\text{increase in } T^{-1}}$$

$$= \frac{-0.65 - 1.85}{(1.70 - 1.50) \times 10^{-3} \, \mathrm{K}^{-1}}$$

$$= \frac{-2.50}{0.20 \times 10^{-3} \, \mathrm{K}^{-1}}$$

$$= -1.25 \times 10^{4} \, \mathrm{K}$$

We saw earlier that the gradient is equal to $-\dfrac{E_a}{R}$ so that

$$-\frac{E_a}{R} = -1.25 \times 10^{4} \, \mathrm{K}$$

The negative signs on either side of the equation cancel, and multiplying both sides by the ideal gas constant R gives

$$E_a = 1.25 \times 10^{4} \, \mathrm{K} \times R$$

Substituting the value of R gives

$$E_a = 1.25 \times 10^{4} \, \mathrm{K} \times 8.31 \, \mathrm{J \, K^{-1} \, mol^{-1}}$$

$$= 1.04 \times 10^{5} \, \mathrm{J \, mol^{-1}}$$

As we might expect, the activation energy has units of J mol^{-1}. Since $1\,kJ = 10^3\,J$ and $10^5 = 10^3 \times 10^2$ we also have

$$E_a = 1.04 \times 10^2\ kJ\ mol^{-1}$$

$$= 104\ kJ\ mol$$

Worked Example 5.18

The activation energy for the reaction

$$2\ H_2O_2 \rightleftharpoons 2H_2O + O_2$$

is 48.9 kJ mol^{-1} in the presence of a colloidal platinum catalyst. What is the effect on the rate constant of raising the temperature from 20°C to 30°C?

CHEMICAL BACKGROUND

This is another example of a disproportionation reaction. If traces of certain ions are present, these catalyse the reaction which then occurs rapidly. One such ion is Fe^{3+}, which may alternate with Fe^{2+} while the reaction is taking place.

Hydrogen peroxide can be produced industrially by the partial oxidation of 2-ethylanthraquinol or 2-propanol by air. It can be extracted into water.

Solution to Worked Example 5.18

If we assume that the rate constant has a value of k_1 when the absolute temperature is T_1, and a value of k_2 at temperature T_2, we can write the logarithmic form of the Arrhenius equation for the two cases as:

$$\ln k_1 = \ln A - \frac{E_a}{RT_1} \text{ and } \ln k_2 = \ln A - \frac{E_a}{RT_2}$$

Notice that the pre-exponential factor is constant for a given reaction, so the value of A is the same in both equations. Subtracting the first of these equations from the second gives

$$\ln k_2 - \ln k_1 = \ln A - \frac{E_a}{RT_2} - \left(\ln A - \frac{E_a}{RT_1} \right)$$

$$= \ln A - \ln A + \frac{E_a}{RT_1} - \frac{E_a}{RT_2}$$

$$= \frac{E_a}{RT_1} - \frac{E_a}{RT_2}$$

$$= \frac{E_a}{R} \left(\frac{1}{T_1} - \frac{1}{T_2} \right)$$

Since

$$\ln X - \ln Y = \ln\left(\frac{X}{Y}\right)$$

as we saw in Section 5.5.2 on page 161, this then becomes

$$\ln\left(\frac{k_2}{k_1}\right) = \frac{E_a}{R}\left(\frac{1}{T_1} - \frac{1}{T_2}\right)$$

and from the left-hand side of the equation, we can obtain the ratio $\frac{k_2}{k_1}$ which indicates the change in the rate constant relative to its initial value k_1.

We need to convert our temperature values in °C to absolute temperature values in K. This gives

$$T_1 = (20+273) = 293 \text{ K}$$

and

$$T_2 = (30+273) = 303 \text{ K}$$

so that we can now calculate the quantity enclosed in brackets in the above equation which is

$$\left(\frac{1}{T_1} - \frac{1}{T_2}\right) = \frac{1}{293 \text{ K}} - \frac{1}{303 \text{ l}}$$

Evaluating these reciprocals using a calculator gives

$$\left(\frac{1}{T_1} - \frac{1}{T_2}\right) = 3.41 \times 10^{-3} \text{K}^{-1} - 3.30 \times 10^{-3} \text{K}^{-}$$

$$= (3.41 - 3.30) \times 10^{-3} \text{K}^{-1}$$

$$= 0.11 \times 10^{-3} \text{K}^{-1}$$

$$= 1.1 \times 10^{-4} \text{K}^{-1}$$

Substituting these values into our equation gives

$$\ln\left(\frac{k_2}{k_1}\right) = \frac{E_a}{R}\left(\frac{1}{T_1} - \frac{1}{T_2}\right)$$

$$= \frac{48.9 \text{ kJ mol}^{-1}}{8.314 \text{ J K}^{-1} \text{mol}^{-1}} \times 1.1 \times 10^{-4} \text{ K}^{-1}$$

$$= \frac{48.9 \times 10^3 \text{ J mol}^{-1}}{8.314 \text{ J K}^{-1} \text{mol}^{-1}} \times 1.1 \times 10^{-4} \text{ K}^{-1}$$

$$= 0.647$$

Notice that the units cancel to give a value for $\ln\left(\dfrac{k_2}{k_1}\right)$ without units, as required. Since

$$\ln\left(\frac{k_2}{k_1}\right) = 0.647$$

we need to take the exponential function of both sides of this equation to remove the logarithmic term as explained in Section 5.7.2 on page 185 and give

$$\frac{k_2}{k_1} = e^{0.647}$$

and with the use of a calculator we obtain

$$\frac{k_2}{k_1} = 1.91$$

We see that for this reaction an increase in temperature of 10°C produces almost a doubling of the rate constant.

This reaction has an activation energy which is fairly typical for reactions which proceed at a reasonable rate. A useful rule of thumb is that a temperature increase of 10°C will produce a doubling of the reaction rate.

5.8 THE STEADY STATE APPROXIMATION

This approximation provides us with a useful means of deriving rate equations for reactions which proceed via the formation of one or more intermediate species. The concentration c of such a species is assumed to increase rapidly at the start of a reaction, remain relatively constant during most of the reaction, and then decrease rapidly at the end of the reaction. The period when the concentration is constant corresponds to a zero gradient on a graph of concentration c against time t, as shown in Figure 5.7. As we saw in Section 4.8.2 on page 139, this corresponds to a zero value of the derivative, i.e.

$$\frac{dc}{dt} = 0$$

5.8.1 SIMULTANEOUS EQUATIONS

If we have a single equation then the most we can ever hope to achieve is to solve it to give the value of a single variable. If we have more than one variable then the only way we can determine their values is if we have more information. We need at least as many equations as there are unknowns; it is also the case that such equations must be independent of each other.

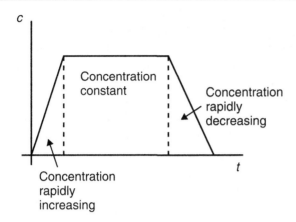

FIGURE 5.7 Schematic graph of the variation in concentration *c* of a reactive intermediate with time *t*.

For example, if we have the equations

$$x - y = 2 \tag{5.1}$$

and

$$2x + 3y = 9 \tag{5.2}$$

which are true at the same time then it follows that the same value of *x* and the same value of *y* must satisfy each one at that time.

It is usual to label simultaneous equations, as we have done here with Eqs. (5.1) and (5.2). Generally, we can solve a pair of simultaneous equations by adding or subtracting them, or by adding or subtracting a multiple of one or both of them, in order to eliminate one of the variables.

In this example, we will choose to eliminate *y*. To do this, we first need to multiply Eq. (5.1) throughout by 3; we will call this Eq. (5.3):

$$3x - 3y = 6 \tag{5.3}$$

We now add Eqs. (5.2) and (5.3). To do this we need to treat each side separately, so we obtain

$$2x + 3y + 3x - 3y = 9 + 6 \tag{5.4}$$

This can be simplified to give

$$5x = 15$$

and dividing both sides by 5 gives

$$x = 3$$

We can now substitute in either of our original equations. If we choose Eq. (5.1) we have

$$3-y=2$$

and adding y to each side of this gives

$$3=2+y$$

from which subtracting 2 from either side gives

$$y=1$$

As a check, we can substitute these values of x and y into the second Eq. (5.2). On the left-hand side of this, we have

$$2x+3y=(2\times3)+(3\times1)=6+3=9$$

which is the required value on the right-hand side.

Worked Example 5.19

In the reaction

$$C_2H_6 \rightarrow C_2H_4+H_2$$

which gives ethene from ethane the intermediates are the radicals CH_3, C_2H_5 and H. Applying the steady state approximation to these gives the simultaneous equations

$$\frac{d[CH_3]}{dt} = 2k_1[C_2H_6]-k_2[CH_3][C_2H_6]=0 \qquad (5.5)$$

$$\frac{d[C_2H_5]}{dt} = k_2[CH_3][C_2H_6]-k_3[C_2H_5]+k_4[H][C_2H_6]-2k_5[C_2H_5]^2=0$$
$$(5.6)$$

$$\frac{d[H]}{dt} = k_3[C_2H_5]-k_4[H][C_2H_6]=0 \qquad (5.7)$$

Solve these equations to give an expression for $[C_2H_5]$ in terms of $[C_2H_6]$.

CHEMICAL BACKGROUND

The above equations are based on the fact that this reaction proceeds by means of a chain reaction. It begins with an initiation step involving the formation of radicals. This can happen in two ways, represented by the equations

$$C_2H_6 \rightarrow 2\ CH_3$$

and

$$CH_3 + C_2H_6 \rightarrow CH_4 + C_2H_5$$

for which the rate constants are denoted as k_1 and k_2 respectively. The propagation steps

$$C_2H_5 \rightarrow C_2H_4 + H$$

and

$$H + C_2H_6 \rightarrow H_2 + C_2H_5$$

can continue in a cyclical fashion to give the products of the reaction. The rate constants of these steps are k_3 and k_4 respectively. The final step is that of termination:

$$2\,C_2H_5 \rightarrow C_4H_{10}$$

for which the rate constant is k_5.

Solution to Worked Example 5.19

Since we require an answer in terms of $[C_2H_5]$ and $[C_2H_6]$ we need to eliminate the terms $[CH_3]$ and $[H]$ from the scheme of equations above.

We can eliminate $[CH_3]$ by adding Eqs. (5.5) and (5.6), and we can eliminate $[H]$ by adding Eqs. (5.6) and (5.7). Consequently, we can eliminate both $[CH_3]$ and $[H]$ by adding Eqs. (5.5), (5.6) and (5.7). This gives

$$2\,k_1\,[C_2H_6] - 2\,k_5\,[C_2H_5]^2 = 0$$

Adding $2\,k_5\,[C_2H_5]^2$ to both sides gives the following expression:

$$2\,k_1\,[C_2H_6] = 2\,k_5\,[C_2H_5]^2$$

Dividing both sides by $2\,k_5$ gives

$$[C_2H_5]^2 = \frac{k_1[C_2H_6]}{k_5}$$

Taking the square root of each side then gives the final answer:

$$[C_2H_5] = \sqrt{\frac{k_1[C_2H_6]}{k_5}}$$

Such calculations are not difficult to perform but do require accuracy in transcribing chemical symbols, particularly of species whose formulae are very similar.

EXERCISES

1. The variable y has the value 17.2 when x is 3.4, and the value 29.7 when x is 10.1. What is the average rate of change of y with respect to x?

2. Determine the following indefinite integrals:

   a. $\int x^5 dx$ b. $\int 7x^6 dx$ c. $\int x^{-4} dx$ d. $\int 2x^{-2} dx$

3. Determine these indefinite integrals.

   a. $\int (6x^2 + 9x + 8) dx$

   b. $\int (3x^3 + 4x^2) dx$

   c. $\int (3x^4 - 4x^2 + 6) dx$

   d. $\int (2x^5 - 4x^3 + x + 7) dx$

4. Calculate these definite integrals.

   a. $\int_{-1}^{1} (x^3 + x^2 + x) dx$

   b. $\int_{0}^{3} (4x^2 + 2x + 1) dx$

   c. $\int_{0}^{1} (4x^5 - 2x^3 + x) dx$

   d. $\int_{-2}^{4} (4x^7 - 6x^5 - 9x^3) dx$

5. Determine these integrals.

   a. $\int \left(x + \dfrac{1}{x} \right) dx$

   b. $\int_{1}^{2} \left(2x^2 + \dfrac{3}{x} \right) dx$

   c. $\int_{3}^{6} \left(8x^3 - \dfrac{6}{x} \right)$

   d. $\int_{-2}^{1} \left(2x^2 + 3x + \dfrac{4}{x^2} \right) dx$

6. Determine:

   a. $\int \dfrac{dx}{x^3}$

   b. $\int_{1}^{5} \dfrac{4}{x^6} dx$

c. $\displaystyle\int_{-3}^{5} \frac{dx}{x^8}$

d. $\displaystyle\int_{-2}^{-1} \frac{dx}{3x^2}$

7. Determine:

a. $\displaystyle\int \left(3x^2 + \frac{5}{x} + \frac{2}{x^3} \right) dx$

b. $\displaystyle\int_{2}^{10} \left(4x^5 - \frac{3}{(x+2)} + \frac{4}{x^5} \right) dx$

c. $\displaystyle\int \left(9x^7 - \frac{4}{(3x-1)} - \frac{3}{x^6} \right) dx$

d. $\displaystyle\int_{1}^{3} \left(2x^3 + \frac{3}{(2x+3)} - \frac{5}{4x^2} \right)$

8. Simplify each of these expressions into a single logarithm:
   a. $\log 3 + \log 4$
   b. $\ln 1 + \ln 2 + \ln 3$
   c. $2 \ln 4 - 3 \ln 2$
   d. $4 \log 6 + 2 \log 3 - \log 9$

9. Express

$$\frac{x}{(x+2)(x+3)}$$

in terms of its partial fractions.

10. Differentiate these expressions.
    a. $2x^2 + 3 \ln x$
    b. $3x + \ln(x^2 - 2)$
    c. $4x^3 - \ln(2x^2 + 1)$
    d. $\ln (4x+3) + \ln (3x-2)$

11. Integrate these expressions.

a. $\displaystyle\int \left(x + \frac{1}{x-2} \right) dx$

b. $\displaystyle\int \left(2x^2 + \frac{1}{x-5} \right) dx$

c. $\displaystyle\int_{3}^{7} \left(\frac{2}{3x+1} - \frac{7}{4x-3} \right) dx$

d. $\displaystyle\int_{0}^{4} \left(\frac{1}{(x+1)} - \frac{3}{(2x+5)} \right) dx$

12. Evaluate the integral

$$\int_{5}^{6} \frac{dx}{(x-4)(x+5)}$$

13. Calculate these quantities:
    a. $2e^{-3}$
    b. $4e^{2}$
    c. $3e^{0}$
    d. $\dfrac{3}{e^{2}} - \dfrac{2}{e^{3}}$

14. Simplify the following expressions:
    a. $(e^{2})^{3}$ b. $e^{4}e^{5}$ c. $\dfrac{x^{2}e^{6}}{e^{3}}$ d. $\dfrac{x^{3}e^{5}}{xe^{4}}$

15. Solve each of the following equations for x:
    a. $10^{x}=42$
    b. $e^{x}=0.75$
    c. $e^{2x-1}=3.2$
    d. $2e^{3x+1}=7.4$

16. Solve each of the following equations for x:
    a. $\ln x=3.6$ b. $\ln 4x=7.2$ c. $\log(2x+3)=6.1$
    d. $\log(4x-1)=0.86$

17. Determine the inverses of these functions.
    a. $f(x)=3x^{2}+5$
    b. $g(x)=2\sqrt{x-8}$
    c. $F(x)=2+\dfrac{7}{8x^{2}}$
    d. $G(x)=\ln(3x^{2}+2)$

18. Solve the simultaneous equations

$$4x+2y=5$$

$$3x+y=9$$

19. Solve the simultaneous equations

$$3x+4y=6$$

$$x-2y=-8$$

20. Solve the simultaneous equations

$$3x+2y+5z=2$$

$$5x+3y-2z=4$$

$$2x-5y-3z=14$$

PROBLEMS

1. Azomethane at an initial pressure of 0.080 mm Hg was allowed to decompose at 350°C. After an hour, its partial pressure was 0.015 mm Hg. What is the average rate of the reaction?

2. Trichloroamine reacts with liquid or concentrated aqueous HCl according to the equation

$$NCl_3 + 4HCl \rightarrow NH_4Cl + 3Cl_2$$

Write down a series of relationships between the rates of change of the concentrations of each of the four chemical species.

3. For a reaction involving reactant A having a rate defined by the equation

$$-\frac{d[A]}{dt} = k[A]^a$$

with k being the rate constant and a some number which may or may not be an integer, obtain the integrated rate equation for the concentration $[A]$ after time t in terms of $[A]_o$, the value when $t=0$. Are there any values of a for which this equation is not valid?

4. It is reasonable to suppose that the rate v of the reaction

$$2\,NO_{(g)} + Cl_{2(g)} \rightarrow 2\,NOCl_{(g)}$$

is given by an equation of the form

$$v = k\,[NO]^m[Cl_2]^n$$

where k is the rate constant. Obtain the logarithmic form of this equation and explain how this could be used to obtain the values of m and n.

5. The concentration of glucose in aqueous hydrochloric acid was monitored and found to give the following results:

$\dfrac{t}{min}$	$\dfrac{c}{10^2\,mol\,dm^{-3}}$
120	5.94
240	5.65
360	5.43
480	5.15

Determine the order of the reaction and the rate constant.

6. Determine an expression for the half-life $t_{\frac{1}{2}}$ of a second-order reaction in which the initial concentrations of the reactants are equal in terms of the rate constant k and the initial concentration of the reactants $[A]_o$.

7. The decomposition of aqueous hydrogen peroxide to gaseous oxygen and liquid water is first-order with a half-life of 6.5 hours. How long does it take for 75% of the starting material to react?

8. At 20°C the rate constant for a reaction is $2.7 \times 10^3 \text{s}^{-1}$. What is the value of the rate constant at 50°C, if the activation energy is 22.3 kJ mol^{-1}?

9. Use the Arrhenius equation to calculate the activation energy for a reaction whose rate constant increases by a factor of 3 when the temperature is increased from 20°C to 30°C.

10. The Eyring equation gives the rate constant k for a reaction in terms of the Boltzmann constant k_B and Planck constant h as

$$k = \frac{k_B T}{h} e^{-\frac{\Delta G^{\ominus}}{RT}}$$

where T is the absolute temperature, R is the gas constant and $\Delta G^{\ominus}$ is the Gibbs energy of activation. Calculate $\Delta G^{\ominus}$ for the reaction

$$H_2 + I_2 \rightarrow 2\,HI$$

for which the value of the rate constant is 0.023 4 dm^{3}mol^{-1}s^{-1} at 400°C.

11. For the reaction between hydrogen and bromine, application of the steady state approximation gives the equations

$$\frac{d[H]}{dt} = k_2[Br][H_2] - k_3[H][Br_2] - k_4[H][HBr] = 0$$

$$\frac{d[Br]}{dt} = 2k_1[Br_2] - k_2[Br][H_2] + k_3[H][Br_2] + k_4[H][HBr] - 2k_5[Br]^2 = 0$$

Solve these equations to give an expression for [Br] in terms of [Br$_2$].

12. For the decomposition of N_2O_5 the following series of equations is obtained:

$$\frac{d[NO_3]}{dt} = k_1[N_2O_5] - k_{-1}[NO_2][NO_3] - k_3[NO][NO_3] = 0$$

$$\frac{d[NO]}{dt} = k_2[NO_2][NO_3] - k_3[NO][NO_3] = 0$$

Solve these equations to give an expression for [NO] in terms of [NO$_2$].

Structural Chemistry

6.1 INTRODUCTION

In this chapter, we will be concerned mainly with the structures of solids, particularly those solids which are crystalline. One of the reasons for this is that far more information is available on the structures of crystalline solids than any other form of matter, this being obtained by the technique of X-ray crystallography.

The concept of symmetry is very important when we are studying structural properties, and we will also look at the use of simple symmetry operators. A more detailed treatment of symmetry involves a reasonable knowledge of group theory, which will not be covered here since an elementary treatment relevant to chemists is usually included in the appropriate section of physical chemistry textbooks.

6.2 PACKING FRACTIONS OF ATOMS IN METALS

We can think of a metal crystal as comprising a series of spherical atoms packed together. There are various ways in which this can occur, and these arrangements give rise to different types of structure. One way in which these differ is in the volume of space actually occupied by atoms; this is known as the packing fraction f_v, and is defined by

$$f_v = \frac{\text{volume occupied by atoms}}{\text{volume of unit cell}}$$

where the unit cell is the unit which defines the structure and which is repeated by symmetry.

Worked Example 6.1

In what is known as a cubic close-packed structure the volume of atoms is determined as $4 \times \frac{4}{3}\pi r^3$ and the volume of the unit cell as $(r\sqrt{8})^3$, where r is the radius of the atom. In an arrangement known as body-centred cubic packing the atoms have volume

DOI: 10.1201/9781003043218-6

$2 \times \dfrac{4}{3} \pi r^3$ and the volume of the unit cell is $\left(\dfrac{4r}{\sqrt{3}} \right)^3$. Calculate the ratio of the packing fractions for the two cases without evaluating π.

CHEMICAL BACKGROUND

The most efficient way of arranging spherical atoms is closest packing. Examples of cubic close-packed structures are calcium, nickel and copper. On the other hand, some alkali metals have a body-centred cubic structure, as do iron and chromium. Figure 6.1 shows the arrangement in the two structures; note that in the body-centred cubic structure layer 3 is a repeat of layer 1.

(a)

(b)

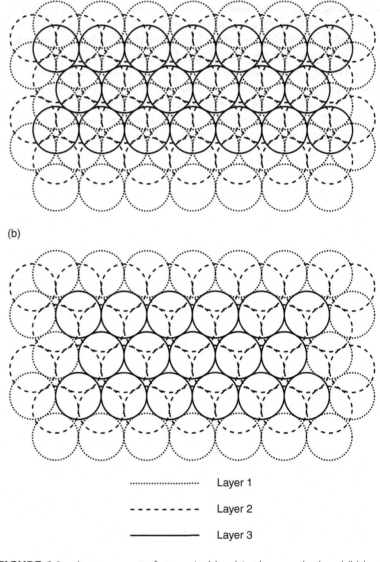

.... Layer 1

– – – – – – – – – Layer 2

———————— Layer 3

FIGURE 6.1   Arrangement of atoms in (a) cubic close-packed and (b) hexagonal close-packed crystals.

Notice that the constant π (3.142...) appears in the expressions because we assume that atoms are spherical. Since there is more empty space in body-centred cubic structures these tend to be more malleable and less brittle.

Solution to Worked Example 6.1

For the cubic close-packed structure, we have

$$f_v(\text{CCP}) = \frac{\text{volume occupied by atoms}}{\text{volume of unit cell}}$$

$$= \frac{4 \times \frac{4}{3}\pi r^3}{(r\sqrt{8})^3}$$

The denominator of this expression can be rewritten as

$$(r\sqrt{8})^3 = r^3 8^{\frac{3}{2}}$$

since using the rules of indices introduced in Section 1.6.3 on page 18

$$\sqrt{8} = 8^{\frac{1}{2}}$$

and

$$(8^{\frac{1}{2}})^3 = 8^{\frac{3}{2}}$$

Performing the multiplication in the numerator then leads to

$$f_v(\text{CCP}) = \frac{\frac{16}{3}\pi r^3}{r^3 8^{\frac{3}{2}}}$$

The term r^3 now cancels to give

$$f_v(\text{CCP}) = \frac{\frac{16}{3}\pi}{8^{\frac{3}{2}}}$$

For the body-centred cubic structure

$$f_v(\text{BCC}) = \frac{\text{volume occupied by atoms}}{\text{volume of unit cell}}$$

$$= \frac{2 \times \frac{4}{3}\pi r^3}{\left(\dfrac{4r}{\sqrt{3}}\right)^3}$$

From Section 1.5.4 on page 15, we recall that dividing by a fraction is the same as multiplying by its reciprocal. Consequently

$$f_v(\text{BCC}) = 2 \times \frac{4}{3}\pi r^3 \times \left(\frac{\sqrt{3}}{4r}\right)^3 = \frac{2 \times 4\pi r^3 \times 3\sqrt{3}}{3 \times 4^3 r^3}$$

$$= \frac{2\pi\sqrt{3}}{4^2}$$

since

$$\left(\sqrt{3}\right)^3 = \sqrt{3} \times \sqrt{3} \times \sqrt{3} = 3\sqrt{3}$$

and several terms cancel top and bottom. We now have

$$\frac{f_v(\text{CCP})}{f_v(\text{BCC})} = \frac{\dfrac{16\pi}{3 \times 8^{\frac{3}{2}}}}{\dfrac{2\pi\sqrt{3}}{4^2}}$$

From Section 1.5.4 on page 15, this gives

$$\frac{f_v(\text{CCP})}{f_v(\text{BCC})} = \frac{16\pi}{3 \times 8^{\frac{3}{2}}} \times \frac{4^2}{2\pi\sqrt{3}} = \frac{16 \times 16}{3 \times 22.62 \times 2 \times 1.732} = \frac{256}{235} = 1.09$$

6.3 ARRANGEMENT OF ATOMS IN CRYSTALS

Crystals are characterized by having atoms in fixed average positions, even though the constant movement of the atoms means that at any instant a particular atom may be slightly displaced. Consequently, it is possible to define interatomic distances with constant values within a crystal.

In the previous section, we met the concept of a unit cell. The unit cell is defined by three edges of lengths a, b and c, and three angles α, β and γ between these edges. The position of an atom within the unit cell can be defined in terms of its position relative to the three edges; these are known as fractional coordinates. It is also possible to define the position of an atom with respect to three axes at right angles, and it is usual to denote these as x, y and z with these quantities being actual rather than fractional lengths.

Here we will use the latter system. It is possible to convert from one to the other but this is a specialized operation which is beyond the scope of this book.

6.3.1 PYTHAGORAS' THEOREM

If we have a right-angled triangle such as the one shown in Figure 6.2 we can identify the two sides of lengths b and c which define the right angle

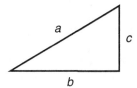

FIGURE 6.2   Definition of lengths in a right-angled triangle.

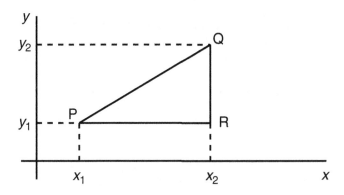

FIGURE 6.3   Right-angled triangle in a two-dimensional coordinate system.

together with the third longer side of length a which is known as the hypotenuse. The lengths of these sides are related by the equation

$$a^2 = b^2 + c^2$$

or, by taking the square root of both sides

$$a = \sqrt{b^2 + c^2}$$

We can now place this triangle within a two-dimensional coordinate system (Figure 6.3), and label the three corners of the triangle as P, Q and R. The point P is assumed to have coordinates (x_1, y_1) and Q is assumed to have coordinates (x_2, y_2). It follows from this that R must have coordinates (x_2, y_1) as shown in Figure 6.3, that length PR is $y_2 - y_1$ and length RQ is $x_2 - x_1$. The distance r between P and Q is then given by Pythagoras' theorem as

$$r = \sqrt{(x_2 - x_1)^2 + (y_2 - y_1)^2}$$

Thus in a planar molecule, if we can assign x and y coordinates to a pair of atoms we can calculate the distance between them.

Worked Example 6.2

A carbon atom in a planar alkene molecule has coordinates (0.250, 0.694), while a hydrogen to which it is bonded has coordinates

(−0.295, 0.250). What is the carbon-hydrogen bond length? The coordinates are given in Å.

CHEMICAL BACKGROUND

A carbon atom in a double carbon-carbon bond occupies a planar environment due to sp^2 hybridization. This ensures that electrons in all the bonds are as far apart as possible, and results in bond angles of 120°. In practice it would not be possible to obtain atomic positions in this molecule from X-ray crystallography; ethane freezes at −183°C and is not a crystalline solid. The unit of Å is 10^{-10} m and is frequently used in structural chemistry as it results in bond length values of around 1–2 Å.

Solution to Worked Example 6.2

Since we have the coordinates of both atoms we can apply Pythagoras' theorem. If we assign

$$x_1 = 0.250 \qquad y_1 = 0.694$$
$$x_2 = -0.295 \qquad y_2 = 0.250$$

We can substitute directly into the equation above to give the bond length r as

$$r = \sqrt{(x_2 - x_1)^2 + (y_2 - y_1)^2}$$

$$= \sqrt{(-0.295 - 0.250)^2 + (0.250 - 0.694)^2}$$

$$= \sqrt{(-0.545)^2 + (-0.444)^2}$$

$$= \sqrt{0.297 + 0.197}$$

$$= \sqrt{0.494}$$

$$= 0.702$$

Note that the answer will be in Å. Strictly speaking, the units should be included throughout the calculation, but here they are omitted for clarity. Notice also that squaring a negative number gives a positive number.

6.3.2 PYTHAGORAS' THEOREM IN THREE DIMENSIONS

Most molecules are not planar, and in such cases, the position of an atom needs to be expressed in terms of three coordinates, usually x, y and z. A direct extension of Pythagoras' theorem then gives the distance r between two atoms having coordinates (x_1, y_1, z_1) and (x_2, y_2, z_2) respectively as

$$r = \sqrt{(x_2 - x_1)^2 + (y_2 - y_1)^2 + (z_2 - z_1)^2}$$

Worked Example 6.3

The coordinates of two atoms in a metal complex are shown below. Determine the distance r between the barium and nitrogen atoms.

	x	y	z
Ba	5.900 0	−3.598 5	6.611 6
N	6.445 2	−4.309 7	9.298 0

CHEMICAL BACKGROUND

X-ray crystallography with single crystals gives the positions of atoms, but no direct information on which atoms are bonded. This has to be inferred from the distance between atoms; bond lengths between specified types of atoms are relatively constant and so bonds can be recognized fairly easily.

Solution to Worked Example 6.3

Substituting in the equation above gives

$$r = \sqrt{(x_2 - x_1)^2 + (y_2 - y_1)^2 + (z_2 - z_1)^2}$$

$$= \sqrt{(6.445\ 2 - 5.900\ 0)^2 + (-4.309\ 7 - (-3.598\ 5))^2 + (9.298\ 0 - 6.611\ 6)^2}$$

$$= \sqrt{0.545\ 2^2 + (-0.711\ 2)^2 + 2.686\ 4^2}$$

$$= \sqrt{0.297\ 2 + 0.505\ 8 + 7.216\ 7}$$

$$= \sqrt{8.019\ 7}$$

$$= 2.832$$

since

$$-4.309\ 7 - (-3.598\ 5) = -4.309\ 7 + 3.598\ 5$$

and the square of a negative number is positive. Once again, the units have been omitted for clarity but are actually Å.

6.4 BRAGG'S LAW

Bragg's Law is mathematically quite simple, and yet the mathematics that underpins it introduces many of the techniques which will be of use to us later on. It governs the behaviour of X-rays when they are diffracted from a crystal.

Since crystals consist of units (atoms, ions or molecules) which are arranged in an orderly fashion, they may be represented by a series of lattice points, each of which has the same environment. Various planes

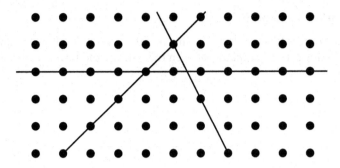

FIGURE 6.4   Planes in a crystal lattice.

may be drawn through these points, as shown in Figure 6.4. Since the spacing of atoms in a crystal is similar to the wavelength of incident X-rays, diffraction rather than reflection occurs with such radiation from planes of atoms. The diffraction angle θ is therefore dependent upon the separation d of particular planes as well as on the wavelength λ of the incident X-rays.

The great advance brought about by the Bragg equation was that it considered radiation of a fixed wavelength. Previously it had only been possible to use 'white' X-rays which were a mixture of various wavelengths and resulted in a far more complicated diffraction pattern.

6.4.1 TRIGONOMETRY

The trigonometric functions are a set of functions which take as their input an angle. The simplest three are defined in terms of a right-angled triangle containing an angle θ, as shown in Figure 6.5. The lengths of the sides of the triangle are labelled as:

- opposite the angle of interest;
- adjacent to the angle of interest;
- the hypotenuse (which is opposite the right angle).

The sine (sin), cosine (cos) and tangent (tan) functions are then defined by the equations

$$\sin \theta = \frac{\text{opposite}}{\text{adjacent}}$$

$$\cos \theta = \frac{\text{adjacent}}{\text{hypotenuse}}$$

$$\tan \theta = \frac{\text{opposite}}{\text{adjacent}}$$

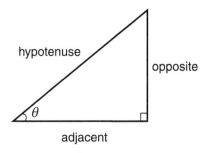

FIGURE 6.5 Right-angled triangle used to define the trigonometric functions.

which show that each function is defined as the ratio of two lengths, and therefore has no units.

While angles are normally expressed in units of degrees, it is sometimes more convenient to express them in terms of the quantity radians, which has the symbol rad. These are related by the equation

$$2\pi\,\mathrm{rad} = 360°$$

so that

$$\pi\,\mathrm{rad} = 180°$$

In practice, it is not usually necessary to convert from one to the other since scientific calculators have both 'degrees' and 'radians' mode. Make sure you are working in the correct one!

The graphs of the sine, cosine and tangent functions are shown in Figure 6.6. It is worth remembering that

$$\sin 0° = \cos 90° = 0$$
$$\sin 90° = \cos 0° = 1$$
$$\tan 0° = \frac{\sin 0°}{\cos 0°} = \frac{0}{1} = 0$$

Also, we will need to be aware that $\cos\theta = \cos(180° - \theta)$ in Section 6.5.3.

Worked Example 6.4

The Bragg equation is expressed as

$$n\lambda = 2d\sin\theta$$

where n is called the order of reflection (having integer values of 1, 2, 3, etc.), and the other terms have been defined above. Calculate the lattice spacing d when copper K_α radiation of wavelength 0.154 nm is incident on a cubic crystal and produces a first-order ($n = 1$) reflection with a scattering angle of 11°.

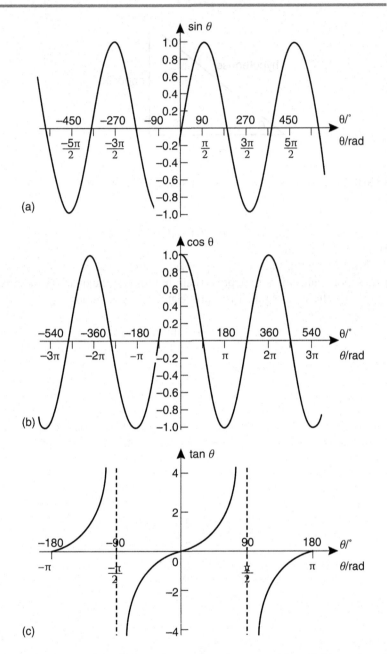

FIGURE 6.6   Graphs of the trigonometric functions: (a) sin θ; (b) cos θ; and (c) tan θ.

CHEMICAL BACKGROUND

X-rays can be generated by firing high-energy electrons at a target, which is most frequently copper metal. Copper electrons are displaced from the innermost K shell and replaced by those from shells further out such as L and M which release their excess energy as X-rays, as shown in Figure 6.7. When replacement of an electron occurs from the next level up L the radiation is known as K_α, while it

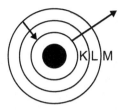

FIGURE 6.7   Production of K_α X-rays.

is also possible (although less likely) to obtain K_β when the replacement electron is from the next highest level M. Other targets used for the generation of X-rays include molybdenum, cobalt, iron and chromium.

Solution to Worked Example 6.4

To calculate the lattice spacing d, we need to rearrange the equation to make d the subject. We can do this by dividing both sides by 2 sin θ:

$$d = \frac{n\lambda}{2 \sin \theta}$$

We have the quantities required to make a straightforward substitution into this equation giving us

$$d = \frac{1 \times 0.154 \text{ nm}}{2 \times \sin 11^\circ}$$

$$= \frac{0.154 \text{ nm}}{2 \times 0.190\,8}$$

$$= 0.404 \text{ nm}$$

Notice that in working through this problem we did not need to convert the units of the wavelength from nm to m, and the units of the spacing of the planes follow directly from substituting into the equation.

6.4.2 INVERSES OF TRIGONOMETRIC FUNCTIONS

The idea of inverse functions was introduced in Section 5.7.2 on page 185. Remember that an inverse function effectively undoes the effect of the original function.

Each of the trigonometric functions we have met has its own inverse. These are

$$\arcsin x \text{ written as } \sin^{-1} x$$

$$\arccos x \text{ written as } \cos^{-1} x$$

$$\arctan x \text{ written as } \tan^{-1} x$$

On a calculator, the inverse trigonometric function is usually obtained by pressing the SHIFT key at the same time as the corresponding trigonometric function. For example, $\sin^{-1}$ is obtained by pressing the SHIFT and sin buttons at the same time.

The use of $^{-1}$ as a superscript in these expressions does *not* have any relationship to its use elsewhere to represent a reciprocal and is purely the nomenclature used to denote the inverse function.

To see how the inverse trigonometric functions are used in practice, consider the function defined as

$$f(x) = \sin(3x + 4)$$

When calculating the inverse of this function, we clearly need to generate x from sin x and this can only be done by using the inverse trigonometric function. We must therefore have the term $\sin^{-1} x$ in the expression for the inverse function $\mathrm{arc}f(x)$. Having done that, we work through the operations in reverse order. In this case, that involves subtracting 4 and then dividing by 3 so that we obtain

$$\mathrm{arc}f(x) = \frac{(\sin^{-1} x) - 4}{3}$$

Worked Example 6.5

A certain set of lattice planes in potassium nitrate crystals has a spacing of 543 pm. Calculate the first-order ($n = 1$) diffraction angle when copper K_α radiation of wavelength 154 pm is incident on these planes.

CHEMICAL BACKGROUND

The crystal structure of potassium nitrate consists of an orthorhombic lattice in which the angles are each 90°. The dimensions of the sides of the unit cell are 5.431, 9.164 and 5.414 Å respectively. Potassium nitrate can be prepared by fractional crystallization from a solution of sodium nitrate and potassium chloride. It is used in gunpowder.

Solution to Worked Example 6.5

It is probably easiest to first determine the value of $\sin \theta$ and then to generate θ by using the inverse function. The first stage is to rearrange the Bragg equation to make $\sin \theta$ the subject. This can be done by dividing both sides by $2d$:

$$\sin \theta = \frac{n\lambda}{2d}$$

which on substitution of the values given leads to

$$\sin \theta = \frac{1 \times 154 \text{ pm}}{2 \times 543 \text{ pm}}$$

$$= 0.1418$$

We now need to apply the inverse sine function to both sides of this equation.

$$\sin^{-1}(\sin \theta) = \sin^{-1}(0.141\ 8)$$

Notice that $\sin^{-1}(\sin \theta) = \theta$. Using a calculator for $\sin^{-1}(0.141\ 8)$ we find

$$\theta = 8.15°$$

This is an example where your calculator needs to be in the correct mode to give the answer in degrees as required.

6.5 THE UNIT CELL

The lattice points referred to in Section 6.4 on page 207 can be used to define the unit cell of a crystal. This is a three-dimensional unit from which it is possible to generate the whole lattice (Figure 6.4). Unit cells may be primitive, having lattice points only at their corners, or non-primitive in which case additional lattice points are present.

6.5.1 UNIT VECTORS

Vectors are numbers which are associated with a particular direction. Because of this, they are very useful for dealing with problems in three dimensions. Any vector can be defined in terms of its components in three directions, normally chosen to be mutually at right angles and denoted as x, y and z. **Unit vectors,** having a size of one unit, can be defined in each of the directions and are denoted as **i, j** and **k** respectively. Note the use of bold typescript to denote vector quantities; when written by hand they can simply be underlined. These unit vectors can be combined by simple addition, so that a vector **a** might be defined as

$$\mathbf{a} = 2\mathbf{i} + 3\mathbf{j} + \mathbf{k}$$

This vector is shown relative to the three axes x, y and z in Figure 6.8.

This defines a vector which can be thought of as starting at the origin, and whose finish is found by moving 2 units in the x-direction, 3 units in the y-direction and 1 unit in the z-direction. The actual vector is then represented diagrammatically by a straight line starting at the origin and finishing at this second point, as in Figure 6.8.

6.5.2 ADDITION AND SUBTRACTION OF VECTORS

Vectors can be added and subtracted using the standard rules of arithmetic applied to each direction separately. For example, if vectors **b** and **c** are defined as

$$\mathbf{b} = 2\mathbf{i} + 3\mathbf{j}$$

$$\mathbf{c} = 2\mathbf{j} + \mathbf{k}$$

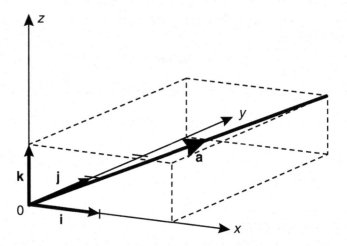

FIGURE 6.8 The vector $2\mathbf{i}+3\mathbf{j}+\mathbf{k}$.

then

$$\mathbf{b}+\mathbf{c}=2\mathbf{i}+3\mathbf{j}+2\mathbf{j}+\mathbf{k}$$

$$=2\mathbf{i}+(3\mathbf{j}+2\mathbf{j})+\mathbf{k}$$

$$=2\mathbf{i}+5\mathbf{j}+\mathbf{k}$$

Worked Example 6.6

Figure 6.9 shows lattice points O and A-G situated at the corners of a primitive orthogonal lattice having dimensions a, b and c; the origin O and axes x, y and z are shown. In terms of the unit vectors $\mathbf{i}$, $\mathbf{j}$ and $\mathbf{k}$, determine the position vector of each point A-G.

CHEMICAL BACKGROUND

Such a crystal lattice, containing points only at the corners of each unit cell, is known as a primitive lattice. Because it has the form of a cuboid if a, b and c are different, it is known as orthorhombic. An example of a structure having a primitive orthorhombic unit cell is that of the mineral Forsterite, $Mg_2(SiO_4)$.

Solution to Worked Example 6.6

We need to consider each lattice point in turn and to define the path which takes us from the origin to that point.

a. To move from O to A, we need to go a units in the x-direction. Since the unit vector in this direction is $\mathbf{i}$, the vector we require is $a\mathbf{i}$.

b. Point B lies on the y-axis at a distance of b units from O. The unit vector along this axis is $\mathbf{j}$, so the required vector is $b\mathbf{j}$.

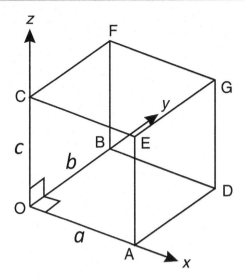

FIGURE 6.9   The primitive orthogonal lattice.

c. Point C lies on the z-axis, along which the unit vector is **k**. We need to move c units, so this point is defined by vector $c\mathbf{k}$.

d. To get from O to D we need to move a units in the x-direction and b units in the y-direction. The vector is consequently $a\mathbf{i}+b\mathbf{j}$.

e. Point E is reached by moving a units in the x-direction and c units in the z-direction. No movement in the y-direction is required. The defining vector is therefore $a\mathbf{i}+c\mathbf{k}$.

f. Point F requires the movement of b units in the y-direction and c units in the z-direction, so the required vector is $b\mathbf{j}+c\mathbf{k}$.

g. Point G is the only one which requires a movement in each of the three directions; a units in x, b units in y and c units in z. The vector defining this point is consequently $a\mathbf{i}+b\mathbf{j}+c\mathbf{k}$.

6.5.3 MULTIPLICATION OF VECTORS

There are two ways in which a pair of vectors can be multiplied together. One results in a pure number, also known as a scalar quantity, while the other gives another vector. The first method is known as calculating the **scalar product,** or dot product since this is the symbol used to denote the process. For example, the scalar product of two vectors **a** and **b** would be denoted as **a** • **b** and is defined as

$$\mathbf{a}\cdot\mathbf{b}=|\mathbf{a}||\mathbf{b}|\cos\theta$$

where $|\mathbf{a}|$ and $|\mathbf{b}|$ are known as the **moduli** of **a** and **b** respectively and θ is the angle between them.

The modulus or magnitude of a vector is its length (a pure number or scalar quantity) and is similar in concept to that met for negative numbers in Sections 4.6.1 (page 130) and 5.6.2 (page 175). It can be calculated by taking the square root of the sums of the squares of the contributions of each unit vector to the total. In two dimensions, this is Pythagoras' Theorem that we met in Section 6.3.1 on page 204. In three dimensions, for example, if

$$\mathbf{v} = 2\mathbf{i} + \mathbf{j} - 3\mathbf{k}$$

then

$$|\mathbf{v}| = \sqrt{(2^2 + 1^2 + (-3)^2}$$
$$= \sqrt{(4 + 1 + 9)}$$
$$= \sqrt{14}$$
$$= 3.7$$

There are two possible angles which can be taken as θ, since $\cos \theta = \cos(180° - \theta)$ and your calculator will offer the acute angle θ. The correct one is shown in Figure 6.10 which shows that both vectors must be pointing away from or both must be pointing towards the point of intersection. The cosine of this angle, known as $\cos \theta$ is easily calculated using a scientific calculator, but it is important to ensure that it has been set to the appropriate mode to accept angles in degrees as input, and that you choose the correct angle θ, or $180° - \theta$. Notice also that $\mathbf{a} \cdot \mathbf{b} = \mathbf{b} \cdot \mathbf{a}$; it does not matter which order you take dot products.

If we consider the scalar products of the various combinations of unit vectors we obtain a very useful set of relationships. For example,

$$\mathbf{i} \cdot \mathbf{i} = |\mathbf{i}||\mathbf{i}| \cos \theta$$

Because $|\mathbf{i}| = 1$ and the angle between $\mathbf{i}$ and $\mathbf{i}$ is zero $\theta = 0°$ and since $\cos 0° = 1$ we then have

$$\mathbf{i} \cdot \mathbf{i} = 1$$

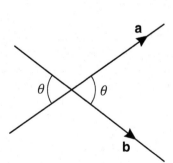

FIGURE 6.10   Definition of the angle θ between two vectors.

The same set of arguments leads to the relationships

$$\mathbf{j} \cdot \mathbf{j} = 1 \text{ and } \mathbf{k} \cdot \mathbf{k} = 1$$

On the other hand, if we consider a scalar product such as $\mathbf{i} \cdot \mathbf{j}$ we then have

$$\mathbf{i} \cdot \mathbf{j} = |\mathbf{i}||\mathbf{j}| \cos \theta$$

where again $|\mathbf{i}| = 1$ and $|\mathbf{j}| = 1$ but this time the angle between $\mathbf{i}$ and $\mathbf{j}$ is a right angle so $\theta = 90°$. Since $\cos 90° = 0$, we now have $\mathbf{i} \cdot \mathbf{j} = 0$ Notice then that

$$\mathbf{i} \cdot \mathbf{j} = \mathbf{j} \cdot \mathbf{i} = 0$$

Similarly

$$\mathbf{j} \cdot \mathbf{k} = \mathbf{k} \cdot \mathbf{j} = 0$$

and

$$\mathbf{i} \cdot \mathbf{k} = \mathbf{k} \cdot \mathbf{i} = 0$$

Now consider what happens if we need to calculate the scalar product of two vectors which are defined in terms of the unit vectors $\mathbf{i}, \mathbf{j}$ and $\mathbf{k}$, such as

$$\mathbf{a} = 3\mathbf{i} + 2\mathbf{j} + \mathbf{k} \text{ and } \mathbf{b} = 2\mathbf{i} - \mathbf{j} + \mathbf{k}$$

We can expand this product as we would with any pair of brackets:

$$\mathbf{a} \cdot \mathbf{b} = (3\mathbf{i} + 2\mathbf{j} + \mathbf{k}) \cdot (2\mathbf{i} - \mathbf{j} + \mathbf{k})$$
$$= 3\mathbf{i} \cdot (2\mathbf{i} - \mathbf{j} + \mathbf{k}) + 2\mathbf{j} \cdot (2\mathbf{i} - \mathbf{j} + \mathbf{k}) + \mathbf{k} \cdot (2\mathbf{i} - \mathbf{j} + \mathbf{k})$$
$$= 6\mathbf{i} \cdot \mathbf{i} - 3\mathbf{i} \cdot \mathbf{j} + 3\mathbf{i} \cdot \mathbf{k} + 4\mathbf{j} \cdot \mathbf{i} - 2\mathbf{j} \cdot \mathbf{j} + 2\mathbf{j} \cdot \mathbf{k} + 2\mathbf{k} \cdot \mathbf{i} - \mathbf{k} \cdot \mathbf{j} + \mathbf{k} \cdot \mathbf{k}$$

Since

$$\mathbf{i} \cdot \mathbf{j} = \mathbf{j} \cdot \mathbf{i} = \mathbf{j} \cdot \mathbf{k} = \mathbf{k} \cdot \mathbf{j} = \mathbf{i} \cdot \mathbf{k} = \mathbf{k} \cdot \mathbf{i} = 0$$

and

$$\mathbf{i} \cdot \mathbf{i} = \mathbf{j} \cdot \mathbf{j} = \mathbf{k} \cdot \mathbf{k} = 1$$

this leaves us with

$$\mathbf{a} \cdot \mathbf{b} = (6 \times 1) - (3 \times 0) + (3 \times 0) + (4 \times 0) - (2 \times 1) + (2 \times 0) + (2 \times 0) + (2 \times 0) - 0 + 1$$

$$= 6 - 2 + 1$$

$$= 5$$

As well as calculating the scalar product of a pair of vectors, there are occasions when we need to multiply vectors and retain some of the information about direction. To do this, we can calculate the **vector product** of a pair of vectors. This is denoted as $\mathbf{a} \times \mathbf{b}$, and is defined as

$$\mathbf{a} \times \mathbf{b} = |\mathbf{a}| \cdot |\mathbf{b}| \cdot \sin\theta \, \hat{\mathbf{n}}$$

where the symbols have the same meaning as in the case of the scalar product. Notice this time, however, that the vector product is defined in terms of a vector $\hat{\mathbf{n}}$.

This gives the direction of the resulting vector, and is itself of magnitude 1. $\hat{\mathbf{n}}$ is perpendicular to both vectors $\mathbf{a}$ and $\mathbf{b}$, and its direction (up or down) can be found by using the corkscrew rule. If we imagine a right-handed corkscrew which rotates from vector $\mathbf{a}$ to vector $\mathbf{b}$, then the direction of travel of the corkscrew gives the direction of the vector $\hat{\mathbf{n}}$.

In Figure 6.10 the vector $\hat{\mathbf{n}}$ will have the direction into the paper using this rule.

Note that

$$\mathbf{a} \times \mathbf{b} \neq \mathbf{b} \times \mathbf{a}$$

because the corkscrew rule results in the vector product pointing in opposite directions.

However

$$\mathbf{a} \times \mathbf{b} = -\mathbf{b} \times \mathbf{a}$$

We can apply this rule to calculate the vector products between pairs of the unit vectors $\mathbf{i}, \mathbf{j}$ and $\mathbf{k}$. Since the angle between pairs of unit vectors is $0°$, $\sin\theta = 0$, and so the vector products between a pair of identical vectors must be zero. This gives

$$\mathbf{i} \times \mathbf{i} = \mathbf{j} \times \mathbf{j} = \mathbf{k} \times \mathbf{k} = 0$$

where the zero is technically a vector quantity. We also know that $\sin 90°$ is 1, so for a pair of perpendicular vectors we do not need to worry about this term and we again obtain a unit vector. If $\mathbf{i}, \mathbf{j}$ and $\mathbf{k}$ form a right-handed set, so that the corkscrew rule can be applied, we obtain

$$\mathbf{i} \times \mathbf{j} = \mathbf{k}$$

$$\mathbf{k} \times \mathbf{i} = \mathbf{j}$$

$$\mathbf{j} \times \mathbf{k} = \mathbf{i}$$

Since $\mathbf{a} \times \mathbf{b} = -\mathbf{b} \times \mathbf{a}$, we also have

$$\mathbf{j} \times \mathbf{i} = -\mathbf{k}$$

$$\mathbf{i} \times \mathbf{k} = -\mathbf{j}$$

$$\mathbf{k} \times \mathbf{j} = -\mathbf{i}$$

This information allows us to calculate the vector product of any pair of vectors which are expressed in terms of the unit vectors **i, j** and **k**. Using the example above, where

$$\mathbf{a} = 3\mathbf{i} + 2\mathbf{j} + \mathbf{k}$$

$$\mathbf{b} = 2\mathbf{i} - \mathbf{j} + \mathbf{k}$$

the expansion of the brackets gives

$$\mathbf{a} \times \mathbf{b} = (3\mathbf{i} + 2\mathbf{j} + \mathbf{k}) \times (2\mathbf{i} - \mathbf{j} + \mathbf{k})$$

$$= 3\mathbf{i} \times (2\mathbf{i} - \mathbf{j} + \mathbf{k}) + 2\mathbf{j} \times (2\mathbf{i} - \mathbf{j} + \mathbf{k}) + \mathbf{k} \times (2\mathbf{i} - \mathbf{j} + \mathbf{k})$$

$$= (3\mathbf{i} \times 2\mathbf{i}) - (3\mathbf{i} \times \mathbf{j}) + (3\mathbf{i} \times \mathbf{k}) + (2\mathbf{j} \times 2\mathbf{i}) - (2\mathbf{j} \times \mathbf{j})$$
$$+ (2\mathbf{j} \times \mathbf{k}) + (2\mathbf{k} \times \mathbf{i}) - (\mathbf{k} \times \mathbf{j}) + (\mathbf{k} \times \mathbf{k})$$

Remembering that

$$\mathbf{i} \times \mathbf{i} = \mathbf{j} \times \mathbf{j} = \mathbf{k} \times \mathbf{k} = 0$$

leaves

$$\mathbf{a} \times \mathbf{b} = -3\mathbf{i} \times \mathbf{j} + 3\mathbf{i} \times \mathbf{k} + 4\mathbf{j} \times \mathbf{i} + 2\mathbf{j} \times \mathbf{k} + 2\mathbf{k} \times \mathbf{i} - \mathbf{k} \times \mathbf{j}$$

We can now substitute for each of these individual vector products to obtain

$$\mathbf{a} \times \mathbf{b} = -3\mathbf{k} + 3(-\mathbf{j}) + 4(-\mathbf{k}) + 2\mathbf{i} + 2\mathbf{j} - (-\mathbf{i})$$

$$= -3\mathbf{k} - 3\mathbf{j} - 4\mathbf{k} + 2\mathbf{i} + 2\mathbf{j} + \mathbf{i}$$

$$= 3\mathbf{i} - \mathbf{j} - 7\mathbf{k}$$

Worked Example 6.7

The methane molecule may be thought of as containing a carbon atom at the centre of a cuboid together with four hydrogen atoms at alternate comers. Use this information to calculate the bond angle in methane.

CHEMICAL BACKGROUND

Using the Valence Shell Electron Pair Repulsion (VSEPR) model, we would expect the four regions of electron density around the central carbon atom to be as far away from each other as possible. This leads to the arrangement described in the question, which is a regular tetrahedron. Such a structure has been confirmed experimentally by means of electron diffraction.

The tetrahedral structure of methane was also predicted by van't Hoff in 1874, on the basis of the observation that only one isomer of

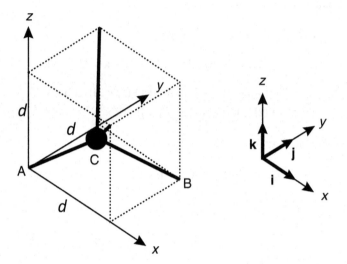

FIGURE 6.11   Structures of CH_3Cl and CH_3Br.

FIGURE 6.12   A carbon atom in a tetrahedral environment.

compounds such as CH_3Cl and CH_3Br has ever been observed, as shown in Figure 6.11.

Solution to Worked Example 6.7

It is necessary to formulate this problem in terms of unit vectors. If we set up the cuboid as shown in Figure 6.12 we can define three axes x, y and z as shown. The carbon atom C is placed at the centre and the four hydrogen atoms at alternate corners. We choose to label two of these A and B and will be calculating the angle A-C-B. Atom A is chosen to be at the origin of our coordinate system, and we will assign the cuboid an arbitrary length d. We are going to determine the angle by calculating the scalar product between vectors $\mathbf{AC}$ and $\mathbf{BC}$.

We can define each of these vectors in terms of the unit vectors $\mathbf{i}$, $\mathbf{j}$ and $\mathbf{k}$ by referring to Figure 6.12. One way of moving from A to C is to move along half the side of the cube in the x-direction, then by the same amount in the y-direction, and finally by the same amount again in the z-direction. This allows us to define the vector from A to C as

$$\mathbf{AC} = \tfrac{1}{2}d\mathbf{i} + \tfrac{1}{2}d\mathbf{j} + \tfrac{1}{2}d\mathbf{k}$$

From Section 6.5.3 on page 215, it should be apparent that both vectors we define must point towards the central carbon atom C in order to give the correct angle. We therefore choose to move from B to C by moving $\frac{1}{2}d$ in the negative x-direction, $\frac{1}{2}d$ in the negative y-direction and $\frac{1}{2}d$ in the positive z-direction. We therefore have

$$\mathbf{BC} = -\tfrac{1}{2}d\mathbf{i} - \tfrac{1}{2}d\mathbf{j} + \tfrac{1}{2}d\mathbf{k}$$

Both of these vectors will have the same modulus, which is

$$|\mathbf{AC}| = |\mathbf{BC}| = \sqrt{\left(\frac{1}{2}d\right)^2 + \left(\frac{1}{2}d\right)^2 + \left(\frac{1}{2}d\right)^2}$$

$$= \sqrt{\frac{d^2}{4} + \frac{d^2}{4} + \frac{d^2}{4}}$$

$$= \sqrt{\frac{3d^2}{4}}$$

$$= d\sqrt{\frac{3}{4}}$$

and so the scalar product is

$$\mathbf{AC} \bullet \mathbf{BC} = d\sqrt{\frac{3}{4}} \times d\sqrt{\frac{3}{4}} \times \cos\theta$$

$$= d^2 \times \frac{3}{4} \times \cos\theta$$

$$= \frac{3}{4} d^2 \cos\theta$$

where θ is the required bond angle.

Calculation of the scalar product by combining the unit vectors and using

$$\mathbf{i} \cdot \mathbf{i} = \mathbf{j} \cdot \mathbf{j} = \mathbf{k} \cdot \mathbf{k} = 1$$

and

$$\mathbf{i} \cdot \mathbf{j} = \mathbf{j} \cdot \mathbf{k} = \mathbf{k} \cdot \mathbf{i} = 0$$

gives

$$\mathbf{AC} \bullet \mathbf{BC} = \left(\frac{1}{2}d\right)\left(-\frac{1}{2}d\right) + \left(\frac{1}{2}d\right)\left(-\frac{1}{2}d\right) + \left(\frac{1}{2}d\right)\left(\frac{1}{2}d\right)$$

$$= -\frac{1}{4}d^2 - \frac{1}{4}d^2 + \frac{1}{4}d^2$$

$$= -\frac{1}{4}d^2$$

These two expressions for the scalar product can now be equated to give

$$\frac{3}{4}d^2 \cos \theta = -\frac{1}{4}d^2$$

Dividing both sides by d^2 and multiplying both sides by 4 gives

$$3 \cos \theta = -1$$

$$\cos \theta = -\frac{1}{3}$$

and so θ can be found by taking the inverse function:

$$\theta = \cos^{-1}\left(-\frac{1}{3}\right)$$

Using a calculator set to 'degrees' mode gives $\theta = 109.5°$ which is the tetrahedral value expected.

Worked Example 6.8

A unit cell can be defined in terms of the three vectors **a, b** and **c** along its edges. The angle between **b** and **c** is known as α, that between **a** and **c** as β and that between **a** and **b** as γ. A monoclinic crystal is defined as one having α and γ equal to 90° with β having a value of greater than 90°. In this case, the vectors **a, b** and **c** are related to the unit vectors **i, j** and **k** by the equations

$$\mathbf{a} = a \sin \beta \mathbf{i} + a \cos \beta \mathbf{k}$$

$$\mathbf{b} = b\mathbf{j}$$

$$\mathbf{c} = c\mathbf{k}$$

where a, b and c are the magnitudes of the vectors **a, b** and **c** respectively. Use the fact that the volume V of the unit cell is given by the triple scalar product

$$V = \mathbf{a} \cdot (\mathbf{b} \times \mathbf{c})$$

to obtain an expression for the volume of the unit cell in terms of a, b, c and β. Note that we actually need to take the modulus $|\mathbf{a} \cdot (\mathbf{b} \times \mathbf{c})|$ of the triple product to account for the fact that it can be negative.

CHEMICAL BACKGROUND

Unit cells can be characterized into various types on the basis of the relationships between the dimensions a, b, c, α, β and γ. The most general case, in which these quantities are all independent, is known as a triclinic unit cell. In the hexagonal cell, a and b are equal, α and β are both 90° and γ is 120°.

Solution to Worked Example 6.8

This question requires us to calculate a **triple scalar product;** it is called this because three vectors are involved and the result will be a scalar quantity.

We evaluate the vector product enclosed in brackets first and then use this result to calculate the scalar product.

The vector product is given by

$$\mathbf{b} \times \mathbf{c} = b\mathbf{j} \times c\mathbf{k}$$

$$= bc \, (\mathbf{j} \times \mathbf{k})$$

since the scalar quantities b and c can be multiplied directly. Since $\mathbf{j} \times \mathbf{k} = \mathbf{i}$, we now have

$$\mathbf{b} \times \mathbf{c} = bc\mathbf{i}$$

We are now able to calculate the required scalar product, which is given by

$$\mathbf{a} \cdot (\mathbf{b} \times \mathbf{c}) = (a \sin \beta \mathbf{i} + a \cos \beta \mathbf{k}) \cdot (bc\mathbf{i})$$

$$= (a \sin \beta \mathbf{i}) \cdot (bc\mathbf{i}) + (a \cos \beta \mathbf{k}) \cdot (bc\mathbf{i})$$

$$= abc \sin \beta \, (\mathbf{i} \cdot \mathbf{i}) + abc \cos \beta \, (\mathbf{k} \cdot \mathbf{i})$$

Since $\mathbf{k} \cdot \mathbf{i} = 0$ and $\mathbf{i} \cdot \mathbf{i} = 1$ this leaves

$$(abc \sin \beta \times 1) + (abc \cos \beta \times 0)$$

so

$$\mathbf{a} \cdot (\mathbf{b} \times \mathbf{c}) = abc \sin \beta$$

which is equal to the volume of the monoclinic unit cell.

6.6 X-RAY DIFFRACTION

Bragg's Law gives us information about the behaviour of diffracted X-rays from one particular set of planes in a crystal. However, as we have seen above, there are many planes existing in a crystal lattice. Interpretation of the complex patterns, arising from the diffraction of X-rays from many different crystal planes simultaneously, enables us to obtain detailed information on the crystal structure. It is generally possible to locate the positions of atoms in a unit cell, and these positions are generally expressed in terms of fractions of each of the dimensions a, b and c, known as fractional crystallographic coordinates and often written as x/a, y/b and z/c.

6.6.1 COMPLEX NUMBERS

A complex number z is one of the forms

$$z = a + ib$$

where a and b are 'real' numbers, which may be positive or negative and have values such as -3.4, 2.6, 7.0 and so on. The number i is defined by the equation

$$i^2 = -1$$

which seems strange at first when you realize that taking the square root of either side leads to

$$i = \sqrt{-1}$$

This i should not be confused with the unit vector $\mathbf{i}$ which we met earlier in this chapter. You may also see the square root of -1 represented by the symbol j, not to be confused with the unit vector $\mathbf{j}$.

In a complex number z, the number a is known as the **real** part of z, and b is known as the **imaginary** part of z. So, if we have the complex number

$$z = 2 - 3i$$

its real part is 2 and its imaginary part -3.

In chemistry, such numbers have been used in crystallography and also in quantum mechanics, as we will see in Chapter 7.

We can perform the usual arithmetic operations on complex numbers. In the case of addition and subtraction, we need to treat the real and imaginary parts separately, so that for example

$$(3 - 2i) + (5 + i) = (3 + 5) + (-2i + i)$$

$$= 8 - i$$

In the case of multiplication, we need to consider every possible term as we have seen previously in Section 3.4.1 on page 64 when multiplying a pair of brackets. For example

$$(2 - 3i)(4 + i) = 2 \times (4 + i) - 3i \times (4 + i)$$

$$= 8 + 2i - 12i - 3i^2$$

Since $i^2 = -1$, this becomes

$$8 - 10i - 3(-1) = 8 - 10i + 3$$

$$= 11 - 10i$$

There is a useful relationship regarding the exponential of an imaginary number, which is a complex number containing only an imaginary part. This is

$$e^{ikx} = \cos kx + i \sin kx$$

$$e^{-ikx} = \cos kx - i \sin kx$$

The exponential function e^x was introduced in Section 5.7.1 on page 185.

Worked Example 6.9

The structure factor $F(h\ k\ l)$ for a particular reflection of X-rays from a crystal is made of contributions from each of the atoms at positions (x_j, y_j, z_j) in a unit cell, and is governed by the values of the integers h, k and l. The expression for its calculation is

$$F(hkl) = \sum f_j \exp 2\pi i(hx_j + ky_j + lz_j)$$

where f_j is the atomic scattering factor for atom j. In a face-centred cubic lattice, the atomic positions (x_j, y_j, z_j) are $(0, 0, 0)$, $(\frac{1}{2}, \frac{1}{2}, 0)$, $(\frac{1}{2}, 0, \frac{1}{2})$ and $(0, \frac{1}{2}, \frac{1}{2})$. Show that if h, k and l are all even or all odd $F(h\ k\ l) = 4f$ but otherwise $F(h\ k\ l) = 0$, assuming all the atoms are identical with atomic scattering factor f.

CHEMICAL BACKGROUND

The structure factor gives the amplitude of all the diffracted waves when the contributions from each atom in the unit cell are added together. The square of this is then proportional to the intensity which can be measured. The atomic scattering factor is a measure of how strongly an atom diffracts and since the electrons are responsible for diffraction, it is proportional to the number of electrons in an atom of a given type.

The integers h, k and l define particular planes in the crystal and, as this problem shows, certain combinations of these values lead to zero structure factors and hence intensities for certain symmetries. These systematic absences, as they are known, are characteristic of particular space groups.

For example, the International Tables for Crystallography lists the reflection conditions for the space group P2$_1$/c as

$$h\,0\,l\ \ l = 2n$$

$$0\,k\,0\ \ k = 2n$$

$$0\,0\,l\ \ l = 2n$$

Thus, for odd values of the indices specified, no reflections will be seen.

Solution to Worked Example 6.9

It may be easiest to begin by assigning labels to each of the four atoms, as shown in the following table:

j	x_j	y_j	z_j
1	0	0	0
2	½	½	0
3	½	0	½
4	0	½	½

so, for example, $z_3 = ½$. The next stage is to expand the expression given for $F(h\,k\,l)$ in terms of the cos and sin functions. This gives

$$F(hkl) = \sum f_j \exp(2\pi i(hx_j + ky_j + lz_j))$$

$$= \sum f_j \left[\cos 2\pi(hx_j + ky_j + lz_j) + i \sin 2\pi(hx_j + ky_j + lz_j) \right]$$

We can expand the summations in this expression to give

$$F(hkl) = f\left[\cos 2\pi\,(hx_1 + ky_1 + lz_1) + i \sin 2\pi\,(hx_1 + ky_1 + lz_1)\right]$$

$$+ f\left[\cos 2\pi\,(hx_2 + ky_2 + lz_2) + i \sin 2\pi\,(hx_2 + ky_2 + lz_2)\right]$$

$$+ f\left[\cos 2\pi\,(hx_3 + ky_3 + lz_3) + i \sin 2\pi\,(hx_3 + ky_3 + lz_3)\right]$$

$$+ f\left[\cos 2\pi\,(hx_4 + ky_4 + lz_4) + i \sin 2\pi\,(hx_4 + ky_4 + lz_4)\right]$$

and then substitute the values of x_j, y_j and z_j for each atom. This gives

$$F(hkl) = f\left[\cos 2\pi\,(h\times0 + k\times0 + l\times0) + i \sin 2\pi\,(h\times0 + k\times0 + l\times0)\right]$$

$$+ f\left[\cos 2\pi\,(h\times½ + k\times½ + l\times0) + i \sin 2\pi\,(h\times½ + k\times½ + l\times0)\right]$$

$$+ f\left[\cos 2\pi\,(h\times½ + k\times0 + l\times½) + i \sin 2\pi\,(h\times0 + k\times½ + l\times½)\right]$$

$$+ f\left[\cos 2\pi\,(h\times0 + k\times½ + l\times½) + i \sin 2\pi\,(h\times0 + k\times½ + l\times½)\right]$$

Removing terms which are obviously zero leaves

$$F(h\,k\,l) = f\left[\cos 0 + i \sin 0\right]$$

$$+ f\left[\cos 2\pi\left(\frac{1}{2}h + \frac{1}{2}k\right) + i \sin 2\pi\left(\frac{1}{2}h + \frac{1}{2}k\right)\right]$$

$$+ f\left[\cos 2\pi\left(\frac{1}{2}h + \frac{1}{2}l\right) + i \sin 2\pi\left(\frac{1}{2}h + \frac{1}{2}l\right)\right]$$

$$+ f\left[\cos 2\pi\left(\frac{1}{2}k + \frac{1}{2}l\right) + i \sin 2\pi\left(\frac{1}{2}k + \frac{1}{2}l\right)\right]$$

This can be simplified since, for example,

$$\cos 2\pi\left(\frac{1}{2}h+\frac{1}{2}k\right)=\cos \pi(h+k)$$

and we know from Section 6.4.1 on page 208 that $\cos 0°=1$ and $\sin 0°=0$, so we obtain

$$F(hkl)=f+f\left[\cos \pi(h+k)+i\sin \pi(h+k)\right]$$

$$+f\left[\cos \pi(h+l)+i\sin \pi(h+l)\right]$$

$$+f\left[\cos \pi(k+l)+i\sin \pi(k+l)\right]$$

If we look at Figure 6.6(a) on page 208 we see that the value of $\sin \theta$ is zero when $\theta=-2\pi, -\pi, 0, \pi, 2\pi$. These are multiples of π, so all of the sine terms in the above equation will be zero since h, k and l are all integers. This leaves us with

$$F(h\ k\ l)=f+f\cos \pi(h+k)+f\cos \pi(h+l)+f\cos \pi(k+l)$$

If we now look at Figure 6.6(b), we see that $\cos \theta$ is 1 when θ has the values $-2\pi, 0$ and 2π, in other words at multiples of 2π. When θ has values of $-\pi$ and π it is equal to -1, in other words at odd multiples of π.

We have four cases to consider. First, if h, k and l are all even then the quantities $\pi(h+k)$, $\pi(h+l)$ and $\pi(k+l)$ will all be even multiples of π, with their respective cosines equal to 1, and so we have

$$F(h\ k\ l)=(f\times 1)+(f\times 1)+(f\times 1)+(f\times 1)=4f$$

Similarly, if h, k and l are all odd, then the quantities $\pi(h+k)$, $\pi(h+l)$ and $\pi(k+l)$ will all be even multiples of π and so, as above,

$$F(h\ k\ l)=4f$$

The situation is more complicated if one or two of h, k and l are odd. For example, if h is odd and k and l are both even then we have

$$h+k \text{ is odd}$$

$$h+l \text{ is odd}$$

$$k+l \text{ is even}$$

Since odd multiples of π give $\cos \theta=-1$ and even multiples of π give $\cos \theta=1$, this will give

$$F(h\ k\ l)=f+f\times(-1)+f\times(-1)+f\times 1$$

$$=f-f-f+f$$

$$=0$$

The same reasoning applies if k or l is odd and the other two indices are even. Now suppose two of h, k and l are odd. If both h and k are odd, we have

$$h + k \text{ is even}$$

$$h + l \text{ is odd}$$

$$k + l \text{ is odd}$$

which leads to

$$F(h\,k\,l) = f + f \times 1 + f \times (-1) + f \times (-1)$$

$$= f + f - f - f$$

$$= 0$$

The same applies to other pairs of odd values of h, k and l.

6.7 SYMMETRY OPERATORS

We have already met the idea that crystals consist of regular repeating units. Another way of expressing this is to say that they possess symmetry. As well as being important in crystals, symmetry may also be evident at the molecular level. For example, when we considered methane in Worked Example 6.7 it was obvious that all the hydrogen atoms were equivalent and that we could choose any pair when calculating the bond angle. If you were to rotate the methane molecule by $120°$ about any bond then you would obtain a molecule which was indistinguishable in orientation from the one you started with. We say that methane has a rotation axis along each bond, but there are several other types of symmetry elements including mirror planes and centres of symmetry.

The subject of symmetry is quite complex, and a thorough understanding requires some knowledge of the branch of mathematics known as group theory. Here we will only look at how individual symmetry operations can be represented.

6.7.1 MATRICES

A matrix is a set of numbers arranged in rows and columns, and enclosed by one pair of brackets per matrix. An example would be

$$\begin{pmatrix} 2 & 3 \\ 4 & 5 \end{pmatrix}$$

This is known as a 2×2 matrix since it has 2 rows and 2 columns. A 2×3 matrix has 2 rows and 3 columns and an example would be

$$\begin{pmatrix} 1 & 2 & 3 \\ 7 & 8 & 9 \end{pmatrix}$$

We can only calculate the sum or difference of two matrices if they have the same dimensions. The operation is performed using the corresponding elements of the two matrices. For example, if

$$\mathbf{A} = \begin{pmatrix} 2 & 3 \\ 4 & 5 \end{pmatrix} \text{ and } \mathbf{B} = \begin{pmatrix} 1 & 2 \\ 3 & 4 \end{pmatrix}$$

we have

$$\mathbf{A} + \mathbf{B} = \begin{pmatrix} 2 & 3 \\ 4 & 5 \end{pmatrix} + \begin{pmatrix} 1 & 2 \\ 3 & 4 \end{pmatrix}$$

$$= \begin{pmatrix} 2+1 & 3+2 \\ 4+3 & 5+4 \end{pmatrix}$$

$$= \begin{pmatrix} 3 & 5 \\ 7 & 9 \end{pmatrix}$$

and

$$\mathbf{A} - \mathbf{B} = \begin{pmatrix} 2 & 3 \\ 4 & 5 \end{pmatrix} - \begin{pmatrix} 1 & 2 \\ 3 & 4 \end{pmatrix}$$

$$= \begin{pmatrix} 2-1 & 3-2 \\ 4-3 & 5-4 \end{pmatrix}$$

$$= \begin{pmatrix} 1 & 1 \\ 1 & 1 \end{pmatrix}$$

It is also possible to multiply a matrix by a number, which effectively acts as a scaling factor. For the example above, we could scale matrix **A** by a factor of 2 so that

$$2\mathbf{A} = 2\begin{pmatrix} 2 & 3 \\ 4 & 5 \end{pmatrix} = \begin{pmatrix} 2\times2 & 2\times3 \\ 2\times4 & 2\times5 \end{pmatrix} = \begin{pmatrix} 4 & 6 \\ 8 & 10 \end{pmatrix}$$

Two matrices can only be multiplied together if the number of columns in the first matrix is the same as the number of rows in the second. It is possible for a 3×2 matrix to be multiplied by a 2×1 matrix; the result would be a 3×1 matrix. However, you cannot multiply a 2×1 matrix by a 3×2 matrix because the number of columns in the first (1) does not match the number of rows in the second matrix (3). The order of multiplying matrices is therefore important.

$$\begin{pmatrix} 2 & 4 \\ 6 & 8 \\ 7 & 1 \end{pmatrix} \begin{pmatrix} 1 & 2 \\ 5 & 6 \end{pmatrix}$$

FIGURE 6.13   The process of matrix multiplication.

To perform the multiplication, we need to multiply each element in the rows of the first matrix by the corresponding elements in the columns of the second matrix. These pairs of numbers are then added to give the element of the matrix which appears in the position given by the intersection of this row and column. For example, if we wish to calculate

$$\begin{pmatrix} 2 & 4 \\ 6 & 8 \\ 7 & 1 \end{pmatrix} \begin{pmatrix} 1 & 2 \\ 5 & 6 \end{pmatrix}$$

then since we are multiplying a 3×2 matrix by a 2×2 matrix we expect to obtain a 3×2 matrix. Matching elements from the first row of the first matrix and the first column of the second matrix as shown in Figure 6.13 gives us

$$(2 \times 1) + (4 \times 5) = 2 + 20 = 22$$

This result appears in the position defined by the intersection, i.e. the *first* row and the *first* column in the results matrix. Now working along the *second* row of the first matrix and the *first* column of the second matrix, we have

$$(6 \times 1) + (8 \times 5) = 6 + 40 = 46$$

and this appears in the position defined by the first column and the second row of the results matrix. We build up the whole matrix in this way to give us the final result:

$$\begin{pmatrix} (2 \times 1) + (4 \times 5) & (2 \times 2) + (4 \times 6) \\ (6 \times 1) + (8 \times 5) & (6 \times 2) + (8 \times 6) \\ (7 \times 1) + (1 \times 5) & (7 \times 2) + (1 \times 6) \end{pmatrix} = \begin{pmatrix} 2 + 20 & 4 + 24 \\ 6 + 40 & 12 + 48 \\ 7 + 5 & 14 + 6 \end{pmatrix}$$

$$= \begin{pmatrix} 22 & 28 \\ 46 & 60 \\ 12 & 20 \end{pmatrix}$$

It is possible to define a matrix for any symmetry operation. For example, the centre of symmetry converts a set of coordinates (x, y, z) into $(-x, -y, -z)$ and could be represented by the matrix

$$\begin{pmatrix} -1 & 0 & 0 \\ 0 & -1 & 0 \\ 0 & 0 & -1 \end{pmatrix}$$

Worked Example 6.10

The operations defined by the symmetry elements in methane can be expressed in matrix notation. The four hydrogen atoms can be represented by the 1×4 matrix

$$(H_A \; H_B \; H_C \; H_D)$$

Multiply this matrix by each of the following 4×4 symmetry matrices to determine the 1×4 matrix which defines the resulting position of each hydrogen atom:

(a) $\begin{pmatrix} 0 & 1 & 0 & 0 \\ 0 & 0 & 1 & 0 \\ 1 & 0 & 0 & 0 \\ 0 & 0 & 0 & 1 \end{pmatrix}$ (b) $\begin{pmatrix} 0 & 0 & 0 & 1 \\ 1 & 0 & 0 & 0 \\ 0 & 1 & 0 & 0 \\ 0 & 0 & 1 & 0 \end{pmatrix}$

(c) $\begin{pmatrix} 0 & 0 & 1 & 0 \\ 0 & 0 & 0 & 1 \\ 0 & 1 & 0 & 0 \\ 1 & 0 & 0 & 0 \end{pmatrix}$ (d) $\begin{pmatrix} 0 & 0 & 1 & 0 \\ 0 & 1 & 0 & 0 \\ 0 & 0 & 0 & 1 \\ 1 & 0 & 0 & 0 \end{pmatrix}$

CHEMICAL BACKGROUND

Methane is said to belong to the point group known as T_d. A point group characterizes the symmetry elements possessed by a molecule; in the case of T_d, these are listed as E, $8C_3$, $3C_2$, $6S_4$, $6\sigma_d$. E is an identity operation which leaves the molecule unchanged. C_3 represents a rotation of $360°/3$ (i.e. $120°$), and C_2 a rotation of $360°/2$ (i.e. $180°$). S_4 is known as an improper rotation and involves rotation of $360°/4$ (i.e. $90°$) followed by reflection across the symmetry plane perpendicular to the axis of rotation. σ_d is a dihedral symmetry plane which bisects the angle formed by a pair of C_2 axes.

Solution to Worked Example 6.10

Application of the first symmetry operator matrix requires us to perform the multiplication

$$\begin{pmatrix} H_A & H_B & H_C & H_D \end{pmatrix} \begin{pmatrix} 0 & 1 & 0 & 0 \\ 0 & 0 & 1 & 0 \\ 1 & 0 & 0 & 0 \\ 0 & 0 & 0 & 1 \end{pmatrix}$$

This is a 1×4 matrix multiplied by a 4×4 matrix and will produce a 1×4 matrix result.

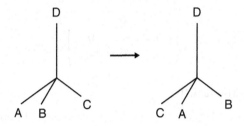

FIGURE 6.14   A three-fold rotation applied to a methane molecule.

We calculate this as shown in the example above, but the presence of several zeros in the symmetry operator matrix assists this process considerably. We will generate the 1×4 matrix by matching elements in the initial 1×4 matrix with elements in each successive column of the 4×4 matrix. For the first column, we obtain

$$(H_A \times 0) + (H_B \times 0) + (H_C \times 1) + (H_D \times 0) = H_C$$

Subsequent columns give

$$(H_A \times 1) + (H_B \times 0) + (H_C \times 0) + (H_D \times 0) = H_A$$

$$(H_A \times 0) + (H_B \times 1) + (H_C \times 0) + (H_D \times 0) = H_B$$

$$(H_A \times 0) + (H_B \times 0) + (H_C \times 0) + (H_D \times 1) = H_D$$

so that the final matrix is

$$(H_C \ H_A \ H_B \ H_D)$$

Comparison with the initial matrix shows that H_A has moved to the position formerly occupied by H_B, H_B to that by H_C, H_C to that by H_A and H_D has remained in the same position. The matrix consequently represents a rotation of 120° about the C-H_D bond, and the net result of performing this symmetry operation is shown in Figure 6.14.

Similar calculations give the following results for applications of the remaining symmetry operator matrices:

$$(b) \ (H_B \ H_C \ H_D \ H_A)$$

$$(c) \ (H_D \ H_C \ H_A \ H_B)$$

$$(d) \ (H_D \ H_B \ H_A \ H_C)$$

Here, (b) and (c) are both S_4 operations which are rotations of 90° followed by reflections through the centre of the molecule, while (d) is a rotation of 120° about the C-H_B bond.

With a little practice, you may be able to see these transformations without needing to perform every step of the calculation.

EXERCISES

1. The two perpendicular sides in a right-angled triangle have lengths of 4.62 and 8.31 cm. What is the length of the hypotenuse?

2. Calculate the distance between the following pairs of points:
   a. (1, 2) and (3, 4)
   b. (7, −2) and (3, 1)
   c. (−6, −5) and (−3, 2)
   d. (−3, −2) and (−1, −4)

3. Calculate the distance between the following pairs of points:
   a. (3, 1, 6) and (4, 3, 4)
   b. (2, 1, −3) and (1, 0, −1)
   c. (5, −9, −2) and (6, −1, 3)
   d. (−3, −2, −7) and (−1, −3, −4)

4. In the right-angled triangle shown in Figure 6.15, calculate $\sin \theta$, $\cos \theta$ and $\tan \theta$, giving your answers correct to 2 decimal places.

5. Determine the angles in the triangle shown in Figure 6.15.

6. Calculate the following, giving your answers correct to three decimal places:
   a. $\sin 11°$
   b. $\cos 44°$
   c. $\cos (−32°)$
   d. $\tan (108°)$

7. Calculate the following to four decimal places:
   e. $\sin (\pi/3)$
   f. $\sin (3\pi/2)$
   g. $\cos (−3\pi/4)$
   h. $\tan (\pi/6)$

8. Solve the following equations, giving your answer in degrees.
   a. $\sin x = 0.352$
   b. $\cos x = 0.406$
   c. $\cos x = −0.108$
   d. $\tan x = 0.764$

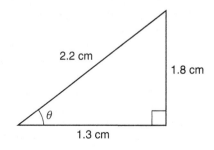

FIGURE 6.15 The right-angled triangle referred to in questions 4 and 5.

9. Solve the following equations, giving your answer in radians.
   a. $\tan (2x+1)=0.472$
   b. $\cos (3x+2)=0.564$
   c. $\sin (x-1)=0.552$
   d. $\sin (2x+3)=0.708$

10. Calculate the modulus of each of these vectors:
   a. $2\mathbf{i}-2\mathbf{j}-3\mathbf{k}$
   b. $2\mathbf{i}+4\mathbf{j}-6\mathbf{k}$
   c. $-3\mathbf{i}+2\mathbf{j}+5\mathbf{k}$
   d. $-\mathbf{i}-4\mathbf{j}-2\mathbf{k}$

11. Determine a unit vector in the direction of each of the following vectors:
   a. $4\mathbf{i}-2\mathbf{j}+6\mathbf{k}$
   b. $-\mathbf{i}-2\mathbf{j}+7\mathbf{k}$
   c. $2\mathbf{i}-6\mathbf{j}+2\mathbf{k}$
   d. $-2\mathbf{i}+3\mathbf{j}-\mathbf{k}$

12. If $\mathbf{a}=\mathbf{i}+2\mathbf{j}-3\mathbf{k}$ and $\mathbf{b}=2\mathbf{i}-\mathbf{j}+\mathbf{k}$, calculate
   a. $\mathbf{a}+\mathbf{b}$
   b. $\mathbf{a}-\mathbf{b}$
   c. $\mathbf{a}\cdot\mathbf{b}$
   d. $\mathbf{a}\times\mathbf{b}$

13. Determine the angle between the vectors $2\mathbf{i}+2\mathbf{j}+\mathbf{k}$ and $\mathbf{i}-\mathbf{j}-2\mathbf{k}$.

14. Determine the real and imaginary parts of each of the following complex numbers:
   a. $4+2i$
   b. $1-8i$
   c. $3-2i$
   d. $-4-7i$

15. Find the values of the following expressions involving complex numbers:
   a. $(3+6i)+(4-3i)$
   b. $(1+6i)+(7-4i)$
   c. $(9-8i)-(7+4i)$
   d. $(-2+4i)-(-3-2i)$

16. Determine the following products of complex numbers:
   a. $(5+2i)(3-i)$
   b. $(1+i)(2+3i)$
   c. $(4-6i)(3+2i)$
   d. $(-2-3i)(-6-i)$

17. Obtain expressions for $\cos kx$ and $\sin kx$ in terms of e^{ikx} and e^{-ikx}.

18. If

$$A = \begin{pmatrix} 2 & 2 \\ 3 & 1 \end{pmatrix} \text{ and } B = \begin{pmatrix} 1 & 0 \\ 4 & 2 \end{pmatrix}$$

calculate
a. $A+B$
b. $A-B$
c. $2A-3B$
d. AB

19. Calculate the product

$$\begin{pmatrix} 3 & 0 & 1 \\ 2 & 4 & 2 \\ 1 & 1 & 3 \end{pmatrix} \begin{pmatrix} 2 \\ 1 \\ 1 \end{pmatrix}$$

20. If

$$A = \begin{pmatrix} 1 & 2 \\ 4 & 2 \end{pmatrix} \text{ and } B = \begin{pmatrix} 2 & 3 \\ 3 & 1 \end{pmatrix}$$

calculate $(2A+B)A$.

PROBLEMS

1. In a simple lattice the volume of atoms is $1 \times \frac{4}{3}\pi r^3$ and the volume of the unit cell is $(2r)^3$. What is the value of the packing fraction?

2. Two carbon atoms in a layer of graphite have coordinates (1.24, 2.62) and (2.09, 3.82) with the distances measured in Å. What is the distance between this pair of atoms?

3. A metal complex has a copper atom at (6.562, 10.888, 4.923) and an oxygen atom at (8.290, 10.352, 4.275). What is the Cu-O bond length, if these measurements are in Å?

4. A nitrogen and a carbon atom in a peptide have respective coordinates (0.840, 0.640, 0.821) and (0.948, 0.601, 0.910), measured in nm. What is the C-N bond length?

5. The second-order reflections from a potassium chloride crystal occur at an angle of 29.2° when radiation of wavelength 1.537 Å is used. Calculate the spacing of the planes which are responsible for this diffraction.

6. The (111) planes in molybdenum carbide have a spacing of 2.47×10^{-10} m. What is the first order diffraction angle when X-rays of wavelength 2.94×10^{-10} m are fired at this crystal?

7. Suppose that the atoms A and B in Figure 6.12 are not equivalent, so that the bond lengths AC and BC are different and

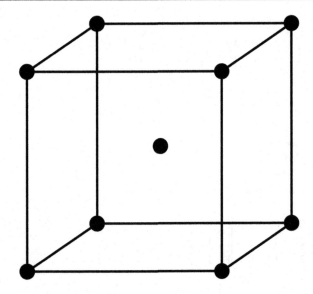

FIGURE 6.16 A body-centred cubic lattice, as referred to in problem 8.

$$\mathbf{AC} = 0.5\,d\,\mathbf{i} + 0.5\,d\,\mathbf{j} + 0.4\,d\,\mathbf{k}$$
$$\mathbf{BC} = -0.5\,d\,\mathbf{i} - 0.5\,d\,\mathbf{j} + 0.6\,d\,\mathbf{k}$$

Calculate the new value of the bond angle.

8. The atoms in a body-centred cubic lattice are situated at the corners of a cube and at its centre, as shown in Figure 6.16. What are the four unique interatomic vectors in this lattice, if the cube has side a?

9. Calculate the volume of the unit cell in the mineral Malachite which has the formula $Cu_2(OH)_2CO_3$ and crystallizes in the space group $P2_1/a$. The unit cell is monoclinic and has dimensions $a = 9.502\,\text{Å}$, $b = 11.974\,\text{Å}$, $c = 3.240\,\text{Å}$ and $\beta = 98.75°$.

10. A crystal containing a 2-fold screw axis has equivalent positions (x, y, z) and $(-x, -y, z+0.5)$.
    a. Obtain a general expression for $F(0\ 0\ l)$.
    b. By considering the positions $(0,0,0)$ and $(0,0,0.5)$ show that $F(0\ 0\ l) \neq 0$ when $l = 2n$.

11. Figure 6.17 shows a benzene molecule, on which has been superimposed x and y-axes. The z-axis is assumed to be perpendicular to the page. The symmetry in the molecule is such that a 60° clockwise rotation moves atom A to B and so on. The matrix $R_z(\theta)$ that defines a rotation of θ about the z-axis is given by:

$$R_z(\theta) = \begin{pmatrix} \cos\theta & -\sin\theta \\ \sin\theta & \cos\theta \end{pmatrix}$$

but by convention here a positive rotation is in the anticlockwise direction.

The coordinates of atom A are given as (1.2, 0.7). By applying successive rotations of 60° generate the coordinates of the remaining atoms B, C, D, E and F.

12. The anti-cancer agent cisplatin has the structure shown in Figure 6.18, which also shows a set of coordinate axes which can be used to define the positions of each atom.

Write a matrix equation for finding the 2×1 matrix

$$\begin{pmatrix} x_2 \\ y_2 \end{pmatrix}$$

which defines the position of N_2 or Cl_2 starting from the corresponding matrix containing x_1 and y_1 which defines the position of the other atom of the same type (N_1 and Cl_1).

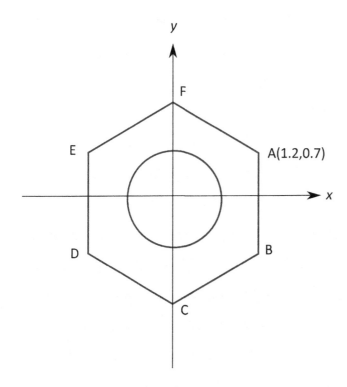

FIGURE 6.17   Benzene molecule referred to in problem 11.

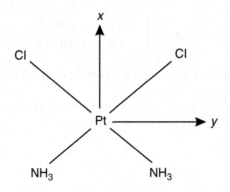

FIGURE 6.18 The structure of cisplatin, referred to in problem 12.

Quantum Mechanics

7.1 INTRODUCTION

The ideas involved in quantum mechanics can be expressed quite concisely, yet a detailed study of individual quantum mechanical systems seems to rely heavily upon mathematics. In some ways, this is unfortunate, as there is a danger that the underlying concepts may be lost in this detail. However, it is essential to be able to apply these ideas if we are to obtain results which are of use and which may be compared with experimental data. In this chapter, we will deal with the mathematics needed for some of the simpler concepts before looking at more complicated situations which need more specialized tools.

7.2 ENERGY LEVEL TRANSITIONS AND APPROPRIATE PRECISION

It is important when performing calculations to quote results to an appropriate number of figures, as discussed in Section 2.2 on page 28. Often we are concerned not to give too many figures. However, in quantum mechanical calculations, there are times when the temptation is to give too few.

Worked Example 7.1

In 1963 the SI unit of length, the metre, was defined as 1 650 763.73 wavelengths of the orange-red radiation from krypton-86. Calculate the (a) wavelength, (b) frequency and (c) energy of this radiation.

CHEMICAL BACKGROUND

Krypton-86 is one of the five stable isotopes of this element. Krypton is a colourless gas which is generally inert, although it can form compounds with fluorine. It finds use in fluorescent lamps.

The current definition of the metre is based on the distance travelled by light in a vacuum during a specified time interval.

Solution to Worked Example 7.1

a. If there are 1 650 763.73 wavelengths in exactly 1 m, the length of a single wavelength will be

DOI: 10.1201/9781003043218-7

$$\frac{1}{1\,650\,763.73}\,m = 6.057\,802\,11 \times 10^{-7}\,m$$

Notice that since we are dealing with a length of exactly 1 m, it is the number of significant figures in the denominator which determines the number of significant figures in the answer.

b. Frequency v is calculated from wavelength λ using the formula

$$v = \frac{c}{\lambda}$$

where c is the velocity of light in a vacuum. In order to avoid losing precision, we need to take the value of c to at least 9 significant figures to match the given data; this is $2.997\,924\,58 \times 10^8\,m\,s^{-1}$. The calculation is then

$$v = \frac{c}{\lambda} = \frac{2.997\,924\,58 \times 10^8\,m\,s^{-1}}{6.057\,802\,11 \times 10^{-7}\,m} = 4.948\,865\,16 \times 10^{14}\,s^{-1}$$

Alternatively, this could be written as $4.948\,865\,16 \times 10^{14}\,Hz$. Note that choosing the value of c to 9 significant figures means that we can also quote v to 9 significant figures.

c. Energy E is given by the formula

$$E = hv$$

where h is the Planck constant. Since v was obtained to 9 significant figures in the previous part, we can ensure that E can be quoted to this level of precision by choosing a value of h to 9 significant figures.

$$h = 6.626\,070\,15 \times 10^{-34}\,J\,s$$

Using this value gives

$$\begin{aligned}
E &= hv \\
&= 6.626\,070\,15 \times 10^{-34}\,J\,s \times 4.948\,865\,16 \times 10^{14}\,s^{-1} \\
&= 3.279\,152\,77 \times 10^{-19}\,J
\end{aligned}$$

quoted to the appropriate level of precision.

7.3 THE PHOTON

The concept of a photon is central to an appreciation of quantum mechanics. It is one unit of electromagnetic radiation, also known as a quantum. This is an example of the idea that quantities traditionally regarded as waves, in this case radiation, can behave as discrete particles.

7.3.1 MATHEMATICAL RELATIONSHIPS

We saw in Section 3.7.4 on page 95 that, if we know that a pair of variables are related proportionally, then it is possible to obtain an expression for the relationship between them. Sometimes, we do not know what this relationship is, or even whether it exists at all. We then need to inspect the data given to see if it is possible to obtain such a relationship.

Worked Example 7.2

Derive an equation which relates the energy E of a photon to its frequency v, given the following data:

$\dfrac{E}{10^{-19}\,\text{J}}$	$\dfrac{v}{10^{14}\,\text{s}^{-1}}$
3.978	6.0
4.973	7.5
6.630	10
9.945	15
19.89	30

Use this equation to calculate the energy corresponding to a frequency of $8.57 \times 10^{14}\,\text{s}^{-1}$.

CHEMICAL BACKGROUND

When light is absorbed by the surface of a metal, electrons are ejected, but this only happens when the frequency of light is above a certain value. Classically, we would expect light of high intensity but low frequency to also cause the ejection of electrons, and this discrepancy led Einstein to propose that light consisted of discrete photons.

This is the basis of the well-known photoelectric effect, which is often used as an introduction to quantum mechanics. The experimental arrangement for this demonstration is shown in Figure 7.1.

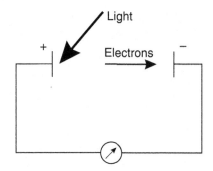

FIGURE 7.1   The apparatus used to illustrate the photoelectric effect.

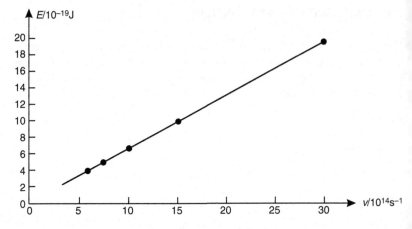

FIGURE 7.2   Graph of photon energy against frequency using the data given in Worked Example 7.2.

The maximum kinetic energy of the electrons is proportional to the difference between the frequency of the incident radiation and the value known as the threshold frequency.

Solution to Worked Example 7.2

Notice that the table headings are given as quotients including the units, as explained in worked Example 3.8 on page 81. Although it would be possible to expand each value in the table in terms of its units, it is actually neater *not* to do this but to work with the table headings as they are given.

A casual inspection of the data shows that as the values of E increase so do the values of v, as shown in Figure 7.2. The simplest relationship we could hope to find, then, would be that of direct proportion. In that case, the ratio E/v would be constant, and equal to the constant of proportionality. If the ratio E/v is constant, it is also true that the ratio

$$\dfrac{E\big/10^{-19}\,J}{v\big/10^{14}\,s^{-1}}$$

$\dfrac{E}{10^{-19}\,J}$	$\dfrac{v}{10^{14}\,s^{-1}}$	$\dfrac{E\big/10^{-19}\,J}{v\big/10^{14}\,s^{-1}}$
3.978	6.0	0.663
4.973	7.5	0.663
6.630	10	0.663
9.945	15	0.663
19.89	30	0.663

is a constant, this being easier to calculate from the data given. Our table of data can then be extended to include this calculation.

The constant ratio term indicates that E and v are indeed directly proportional and are related by the equation

$$\frac{E\big/10^{-19}\,\text{J}}{v\big/10^{14}\,\text{s}^{-1}} = 0.663$$

We can multiply both sides by $v/10^{14}\,\text{s}^{-1}$ so that this becomes

$$\frac{E}{10^{-19}\,\text{J}} = \frac{0.663 \times v}{10^{14}\,\text{s}^{-1}}$$

We can also multiply both sides by $10^{-19}\,\text{J}$ to obtain

$$E = 0.663 \times \frac{v}{10^{14}\,\text{s}^{-1}} \times 10^{-19}\,\text{J}$$

Using the rules of indices from Section 1.6 on page 16 we find that the powers $10^{-19}/10^{14}$ combine to give $10^{(-19-14)}$ or 10^{-33} and the unit J/s^{-1} is the same as J s. This then gives the relationship as

$$E = (0.663 \times 10^{-33}\,\text{J s})v$$
$$= (6.63 \times 10^{-34}\,\text{J s})v$$

The quantity in brackets is known as **Planck's constant** and is normally given the symbol h, so we obtain the relationship

$$E = hv$$

where $h = 6.63 \times 10^{-34}\,\text{J s}$ as seen in the previous section.

We can now substitute the value of v of $8.57 \times 10^{14}\,\text{s}^{-1}$ to give

$$E = 6.63 \times 10^{-34}\,\text{J s} \times 8.57 \times 10^{14}\,\text{s}^{-1}$$
$$= 56.82 \times 10^{-34+14}\,\text{J s s}^{-1}$$
$$= 56.82 \times 10^{-20}\,\text{J}$$
$$= 5.68 \times 10^{-19}\,\text{J}$$

since the units s and s^{-1} cancel. Notice that since the frequency value of $8.57 \times 10^{14}\,\text{s}^{-1}$ falls within the range of the values given in the question ($6.0 - 30 \times 10^{14}\,\text{s}^{-1}$) this is known as interpolation. If we had been calculating the energy by using a frequency value outside this range, it would be known as extrapolation (Figure 7.3). We can be less confident of extrapolating reliably since the relationship we have obtained may not be valid over other ranges of frequency.

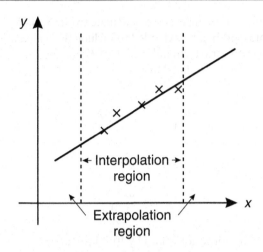

FIGURE 7.3   Interpolation and extrapolation.

7.4 FORCES BETWEEN ATOMS

Despite the successes of quantum mechanics, the classical view of atoms can still be very useful and is more appropriate in some circumstances. We are often interested in the interactions between atoms, as these allow the prediction of properties such as the energy and structure of a system.

7.4.1 PROPORTIONALITY

The subject of proportion was discussed in Section 3.7.4 on page 95. There we saw that if two quantities x and y are proportional to each other then $y/x = k$ or $y = kx$ where k is the constant of proportionality.

Worked Example 7.3

One model of a chemical bond is that of a simple spring, as shown in Figure 7.4, in which the force acting on the spring is proportional to the displacement of the bond length from its equilibrium value. In this case, the constant of proportionality is known as the force constant of the bond and has the value of 440 N m^{-1} for a single bond

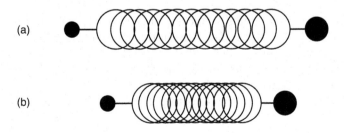

FIGURE 7.4   The (a) displacement of a spring from its (b) equilibrium position.

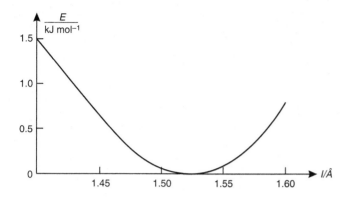

FIGURE 7.5   The energy arising from the displacement of a bond from its equilibrium length in Worked Example 7.3.

between a pair of carbon atoms. Calculate the force on this bond if it is displaced by 2.5 pm from its equilibrium position.

CHEMICAL BACKGROUND

Such calculations form the basis of the molecular mechanics method for calculating molecular structures and energies. The energy E arising from the displacement of a particular bond is calculated from an equation of the form

$$E = \frac{1}{2}k(l - l_o)^2$$

where k is the force constant, l is the observed length and l_o is the equilibrium length. Figure 7.5 shows a plot of E against l for the displacement of the bond being discussed here.

Typical values of these parameters for a single bond between a pair of carbon atoms would be $k = 4.40$ mdyn Å^{-1} and $l_o = 1.523$ Å. These non-SI units are invariably used in calculations of this type.

Solution to Worked Example 7.3

We need to set up an equation to relate the force F to the displacement x. Since $F \propto x$, it follows that

$$F = kx$$

and we are given the value of the proportionality constant k in the question. We substitute the values given into this equation to obtain

$$
\begin{aligned}
F &= 440 \text{ N m}^{-1} \times 2.5 \text{ pm} \\
&= 440 \text{ N m}^{-1} \times 2.5 \times 10^{-12} \text{ m} \\
&= 1\,100 \times 10^{-12} \text{ N m}^{-1} \text{ m} \\
&= 1.1 \times 10^{-9} \text{ N} \\
&= 1.1 \text{ nN}
\end{aligned}
$$

Expressing the answer in nN (10^{-9} N) results in a value which is easily read.

7.4.2 STATIONARY POINTS

Stationary points were discussed in Section 4.8.2 on page 139. We saw that the stationary points of the function $f(x)$ occurred when

$$\frac{df(x)}{dx} = 0$$

and that

$$\frac{d^2 f(x)}{dx^2} < 0$$

denotes a maximum, while

$$\frac{d^2 f(x)}{dx^2} > 0$$

denotes a minimum.

Worked Example 7.4

The energy $E(r)$ of the interaction between two molecules separated by a distance r is given by the function

$$E(r) = -\frac{A}{r^6} + \frac{B}{r^{12}}$$

Locate and identify the stationary points of the function $E(r)$.

The constants A and B are related to the depth ε of the potential well and the equilibrium intermolecular separation r_e. For xenon, these quantities have the values 1.9 kJ mol^{-1} and 4.06 Å, respectively, the potential curve being illustrated in Figure 7.6.

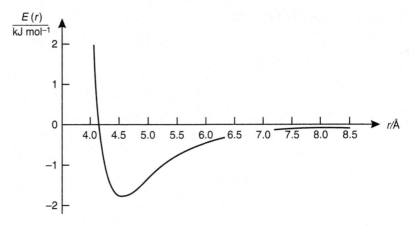

FIGURE 7.6 Graph of the interaction energy $E(r)$ as defined in Worked Example 7.4.

CHEMICAL BACKGROUND

The function given in this worked example represents the interaction energy between a pair of molecules, but such functions have been used as 'interatomic potentials' to model the interactions of atoms in the solid phase. Other types of function can be used to represent these interactions, and parameters such as A and B need to be varied to give the best agreement with experimental data.

Solution to Worked Example 7.4

We saw in Section 4.8.2 that the stationary points of a function $f(x)$ were given by those values of x which make

$$\frac{df(x)}{dx} = 0$$

so in this worked example, we need to find the value of r which gives

$$\frac{dE(r)}{dr} = 0$$

We also saw that the nature of those stationary points is determined by evaluating the sign of

$$\frac{d^2 f(x)}{dx^2}$$

so in this worked example, we need to evaluate

$$\frac{d^2 E(r)}{dr^2}$$

which corresponds to the value of r which gives

$$\frac{dE(r)}{dr} = 0$$

The differentiation of $E(r)$ is easier to perform if we rewrite the function as

$$E(r) = -Ar^{-6} + Br^{-12}$$

since

$$\frac{1}{x^n} = x^{-n}$$

Using the rule that

$$\frac{d}{dx}(ax^n) = anx^{n-1}$$

Section 3.5.1 on page 71 then gives us

$$\frac{d}{dr}(-Ar^{-6}) = (-A)(-6)r^{-6-1}$$

$$= 6Ar^{-7}$$

and

$$\frac{d}{dr}(Br^{-12}) = B \times (-12r^{-12-1})$$

$$= -12Br^{-13}$$

so that

$$\frac{dE(r)}{dr} = \frac{6A}{r^7} - \frac{12B}{r^{13}}$$

This will be zero when

$$\frac{6A}{r^7} = \frac{12B}{r^{13}}$$

This equation can be rearranged by multiplying both sides by r^{13} and dividing both sides by $6A$ to give

$$\frac{r^{13}}{r^7} = \frac{12B}{6A}$$

$$r^{13-7} = \frac{2B}{A}$$

$$r^6 = \frac{2B}{A}$$

We can finally solve this equation by taking the sixth root of either side, since

$$\sqrt[6]{r^6} = r$$

In other words, the process of taking the sixth root is the inverse of raising a number to power 6. The same relationship holds for any other power. Having done this, we finally obtain

$$r = \sqrt[6]{\frac{2B}{A}}$$

This tells us the position of the stationary point but nothing about its nature. We now need to calculate $\dfrac{d^2 E(r)}{dr^2}$ which we can do by differentiating the equation

$$\frac{dE(r)}{dr} = \frac{6A}{r^7} - \frac{12B}{r^{13}}$$

which is easier to do if we rewrite it as

$$\frac{dE(r)}{dr} = 6Ar^{-7} - 12Br^{-13}$$

Using the same rules of differentiation as before gives

$$\frac{d^2E}{dr^2} = (6\times(-7))Ar^{-7-1} + ((-12)\times(-13))Br^{-13-1}$$

$$= -42Ar^{-8} + 156Br^{-14}$$

$$= -\frac{42A}{r^8} + \frac{156B}{r^{14}}$$

Since the minimum occurs when

$$r = \sqrt[6]{\frac{2B}{A}}$$

it follows that

$$r^6 = \frac{2B}{A}.$$

It helps if we can rewrite our expression for $\dfrac{d^2E}{dr^2}$ in terms of powers of r^6.

$$\frac{d^2E}{dr^2} = -\frac{42A}{r^8} + \frac{156B}{r^{14}}$$

$$= \frac{1}{r^2}\left(-\frac{42A}{r^6} + \frac{156B}{r^{12}}\right)$$

If we realize that

$$r^{12} = (r^6)^2$$

we then have

$$r^{12} = \left(\frac{2B}{A}\right)^2$$

$$= \frac{2^2 B^2}{A^2}$$

$$= \frac{4B^2}{A^2}$$

and so

$$\frac{d^2E}{dr^2} = \frac{1}{r^2}\left(-\frac{42A}{2B/A} + \frac{156B}{4B^2/A^2}\right)$$

$$= \frac{1}{r^2}\left(-\frac{42A^2}{2B} + \frac{156BA^2}{4B^2}\right)$$

$$= \frac{1}{r^2}\left(-\frac{21A^2}{B} + \frac{39A^2}{B}\right)$$

$$= \frac{1}{r^2} \times \frac{18A^2}{B}$$

$$= \frac{18A^2}{r^2B}$$

Since A and B are both positive constants $\dfrac{d^2E}{dr^2}$ must be greater than zero, indicating that $r = \sqrt[6]{\dfrac{2B}{A}}$ denotes a minimum.

7.5 PARTICLE IN A BOX

One of the first quantum mechanical systems treated in a study of the subject is usually the particle in a box. This consists of a particle constrained within a one-dimensional box of a specified width, within which the potential energy is zero and outside of which the potential energy is infinite. The model is readily extended to two- and three-dimensional systems.

7.5.1 COMPLEX NUMBERS

We met the concept of complex numbers in Section 6.6.1 on page 224. In particular, we saw that a complex exponential could be expressed by the equation

$$e^{ikx} = \cos kx + i \sin kx$$

Worked Example 7.5

The wavefunctions Ψ for the particle in a one-dimensional box are given by the expression

$$\Psi = \left(\frac{2}{a}\right)^{\frac{1}{2}}\left(\frac{1}{2i}\right)(e^{ikx} - e^{-ikx})$$

where $i^2 = -1$, a is the length of the box, x is the position of the particle within the box and k is a constant. Simplify this expression by removing the imaginary part to leave a real number.

CHEMICAL BACKGROUND

The constant k is actually given by the formula

$$k = \frac{\sqrt{8\pi^2 mE}}{h}$$

where m is the mass of the particle, E is energy and h is Planck's constant. If we apply the boundary conditions that the wavefunction must fall to zero at $x = 0$ and $x = a$, we can solve this expression to give the allowed values of the energy, as we will see in Worked Example 7.6.

Solution to Worked Example 7.5

We need to use the relationships

$$e^{ikx} = \cos kx + i \sin kx$$

and

$$e^{-ikx} = \cos kx - i \sin kx$$

We then have

$$\Psi = \left(\frac{2}{a}\right)^{\frac{1}{2}} \left(\frac{1}{2i}\right)(e^{ikx} - e^{-ikx})$$

$$= \left(\frac{2}{a}\right)^{\frac{1}{2}} \left(\frac{1}{2i}\right)[\cos kx + i \sin kx - (\cos kx - i \sin kx)]$$

$$= \left(\frac{2}{a}\right)^{\frac{1}{2}} \left(\frac{1}{2i}\right)[\cos kx + i \sin kx - \cos kx + i \sin kx]$$

The two terms in $\cos kx$ cancel out and we are left with

$$\Psi = \left(\frac{2}{a}\right)^{\frac{1}{2}} \left(\frac{1}{2i}\right)[2i \sin kx]$$

$$= \left(\frac{2}{a}\right)^{\frac{1}{2}} \sin kx$$

The graph of this wavefunction is shown in Figure 7.7.

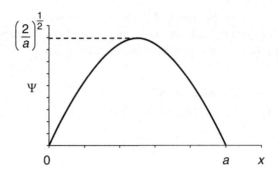

FIGURE 7.7   Graph of the wavefunction ψ for the particle in a one-dimensional box of length a.

7.5.2 SEQUENCES

In Section 3.4.3 on page 69, we met the idea of a function for the first time. This was represented by $f(x)$, where x was a number upon which certain operations (such as multiplication by a constant, raising to a power and so on) were performed. A sequence differs from a function only in that x is replaced by n, where n is an integer (i.e. a whole number) and is therefore restricted to the values 0, 1, 2, 3, 4,....

For example, we could define a sequence as

$$f(n) = 3n + 1 \text{ for } n \geq 0$$

We would then have

$$f(0) = (3 \times 0) + 1 = 1$$

$$f(1) = (3 \times 1) + 1 = 4$$

$$f(2) = (3 \times 2) + 1 = 7$$

$$f(3) = (3 \times 3) + 1 = 10$$

and so on. It is necessary to specify the starting value of n, which will usually be 0 or 1. This can be done using a greater than sign > or a greater than or equal to sign $\geq$. These input values are known as the domain of the sequence.

In quantum mechanics, we are often concerned with the idea that particles may only occupy discrete energy levels which are also labelled with integers (usually from 1 upwards).

Worked Example 7.6

The energy E of each level n in the one-dimensional particle in a box problem is given by the function

$$E(n) = \frac{n^2 h^2}{8ma^2}$$

FIGURE 7.8   Structure of hexatriene.

for $n \geq 1$ where m is the mass of the particle, a is the width of the box and h is Planck's constant. Find the sum of the first five terms of this sequence in terms of a, h and m.

CHEMICAL BACKGROUND

This model would be applicable to the pi electrons in hexatriene which are delocalized and can be thought of as being contained in a one-dimensional box of length 7.3 Å, this being the distance between the terminal carbon atoms, plus half a bond length on either side (Figure 7.8).

Solution to Worked Example 7.6

We need to add together five terms of the form

$$\frac{n^2 h^2}{8ma^2}$$

Once we realize that the quantity $\dfrac{h^2}{8ma^2}$ is common to each of these terms, we can write the required sum as

$$\frac{h^2}{8ma^2}(1^2 + 2^2 + 3^2 + 4^2 + 5^2)$$

$$= \frac{h^2}{8ma^2}(1 + 4 + 9 + 16 + 25)$$

$$= \frac{55h^2}{8ma^2}$$

These energy levels are shown in Figure 7.9.

This technique is analogous to summing the energies of electrons in a molecule where each pair occupies a separate energy level.

7.5.3 INVERSE FUNCTIONS

This topic has been discussed in Section 5.7.2 on page 185. We saw that the effect of an inverse function $\text{arc}f(x)$ is to undo the effect of the original function $f(x)$. In particular

$$f(\text{arc}f(x)) = x \text{ and } \text{arc}f(f(x)) = x$$

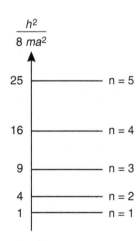

FIGURE 7.9   The energies available to a particle confined to a one-dimensional box.

Worked Example 7.7

The quantum mechanical energy E of a particle in a three-dimensional box is given by the function

$$E(n_x, n_y, n_z) = \frac{h^2}{8m}\left(\frac{n_x^2}{a^2} + \frac{n_y^2}{b^2} + \frac{n_z^2}{c^2}\right)$$

where h is Planck's constant and m the mass of the particle. The quantum numbers n_x, n_y and n_z relate to the directions x, y and z in which the box has dimensions a, b and c respectively.

Obtain an expression for the inverse function $arcE(n)$ in the special case where $n_x = n_y = n_z = n$ and $a = b = c = d$ which is the situation existing in a cube.

CHEMICAL BACKGROUND

The extension of the quantum mechanics of the particle in a one-dimensional box to three dimensions is relatively straightforward. This is due to the fact that the motion in the three perpendicular directions x, y and z can be treated independently. If this was not so, the solution to the problem would be much more difficult to obtain. This ability to partition the energy has important implications in both quantum mechanics and statistical mechanics.

Since we are now dealing with three quantum numbers, it is possible to obtain degenerate energy levels. These have the same energies but different quantum numbers, for example $E(1,2,1)$ and $E(1,1,2)$ when $a = b = c$.

This model could be used to represent, say, nitrogen gas in a 10 cm by 10 cm by 10 cm container. The calculation would then show that the separation between the energy levels is extremely small, and so there is no effective quantisation on this scale.

Solution to Worked Example 7.7

The first stage in this problem is to obtain an expression for $E(n)$, the energy of a particle in a cube. If we make the substitutions

$$n_x = n_y = n_z = n$$

and

$$a = b = c = d$$

into the given equation, we obtain

$$E(n,n,n) = \frac{h^2}{8m}\left(\frac{n^2}{d^2} + \frac{n^2}{d^2} + \frac{n^2}{d^2}\right)$$

$$= \frac{3h^2 n^2}{8md^2}$$

As this is now a function of a single variable, n, it makes sense to write

$$E(n) = \frac{3h^2 n^2}{8md^2}$$

Before we consider the inverse of this function, arc $E(n)$, it is worth considering the stages involved in calculating $E(n)$ from a given value of n. These are:

1. Square n to give n^2
2. Multiply n^2 by $3h^2$ to give $3h^2 n^2$
3. Divide $3h^2 n^2$ by $8md^2$ to give

$$\frac{3h^2 n^2}{8md^2}$$

The reverse of this process is then as follows:

1. Multiply n by $8md^2$ to give $8md^2 n$.

2. Divide $8md^2 n$ by $3h^2$ to give $\dfrac{8md^2 n}{3h^2}$.

3. Take the square root of $\dfrac{8md^2 n}{3h^2}$ to give

$$arcE(n) = \sqrt{\frac{8md^2 n}{3h^2}}$$

7.5.4 DIFFERENTIATION OF FRACTIONAL INDICES

When differentiation was introduced in Section 3.5.1 on page 71, we saw that

$$\frac{d}{dx}(ax^n) = anx^{n-1}$$

In other words, if we differentiate an expression containing a power of x, we multiply by the power and reduce the power by 1. Although the examples of this given so far have all involved integral powers of x, this by no means has to be the case. So, for example

$$\frac{d}{dx}\left(x^{\frac{2}{3}}\right) = \frac{2}{3}x^{\frac{2}{3}-1}$$

$$= \frac{2}{3}x^{-\frac{1}{3}}$$

This could also be written as

$$\frac{2}{3x^{\frac{1}{3}}}$$

since

$$x^{-n} = \frac{1}{x^n}$$

A second example is

$$\frac{d}{dx}(3x^{-0.25}) = 3 \times (-0.25)x^{-0.25-1}$$

$$= -0.75x^{-1.25}$$

Worked Example 7.8

The density of states is the number of quantum states per unit energy at a given energy, and is an important characteristic of a quantum system. For a system whose energy E is given in terms of a single quantum number n, the density of states $\rho(E)$ is given by

$$\rho(E) = \frac{dn}{dE}$$

Determine $\rho(E)$ for the particle in a one-dimensional box.

CHEMICAL BACKGROUND

The density of states function gives us a way of measuring energy level spacing. For some systems, this spacing will be constant while for others it will vary. A further complication occurs in degenerate systems, where the degeneracy has to be included in the function as well as the spacing.

Solution to Worked Example 7.8

In order to calculate $\frac{dn}{dE}$ we need to rearrange our expression for E to make n the subject. Starting from

$$E = \frac{n^2 h^2}{8ma^2}$$

we multiply both sides by $8ma^2$ to give

$$8ma^2 E = n^2 h^2$$

Now divide both sides by h^2 to give

$$\frac{8ma^2 E}{h^2} = n^2$$

Finally we can take the square root of each side to give

$$\frac{a}{h}(8mE)^{\frac{1}{2}} = n$$

We can rewrite this in the opposite order with the variable E separate as

$$n = \frac{a}{h}(8m)^{\frac{1}{2}} E^{\frac{1}{2}}$$

Before differentiating this it may be useful to determine that

$$\frac{d}{dx}(x^{\frac{1}{2}}) = \frac{1}{2}x^{\frac{1}{2}-1} = \frac{1}{2}x^{-\frac{1}{2}}$$

so that

$$\frac{dn}{dE} = \frac{a}{h}(8m)^{\frac{1}{2}}\left(\frac{1}{2}E^{-\frac{1}{2}}\right)$$

Since

$$8^{\frac{1}{2}} = (4\times2)^{\frac{1}{2}} = 4^{\frac{1}{2}} \times 2^{\frac{1}{2}} = 2\times2^{\frac{1}{2}}$$

and

$$E^{-\frac{1}{2}} = \frac{1}{E^{\frac{1}{2}}}$$

We can write the expression for the density of states as

$$\frac{dn}{dE} = \frac{a}{h}(2\times2^{\frac{1}{2}})m^{\frac{1}{2}} \times \frac{1}{2E^{\frac{1}{2}}}$$

$$= \frac{a}{h} \times \frac{2^{\frac{1}{2}}m^{\frac{1}{2}}}{E^{\frac{1}{2}}}$$

so that

$$\rho(E) = \left(\frac{a}{h}\right)\left(\frac{2m}{E}\right)^{\frac{1}{2}}$$

7.5.5 USE OF STANDARD INTEGRALS

In general, the process of integration is more difficult than that of differentiation. As chemists, it therefore makes sense when faced with an integral to evaluate, to see whether the result is available rather than starting from scratch. This is the idea behind many compilations of tabulations of such standard integrals. The trick in using these is to recognize which form is appropriate; frequently this requires a certain degree of manipulation of the function to be integrated.

Suppose that we wish to determine the integral

$$\int_{0.0}^{0.5} \frac{dx}{\sqrt{1-2x^2}}$$

given the standard integral

$$\int \frac{dx}{\sqrt{a^2-x^2}} = \sin^{-1}\left(\frac{x}{a}\right) + C$$

The first step is to recast the function to be integrated into the required form. This can be done by realizing that $2x^2$ needs to be become x^2; this can be done by dividing the square root by 2, and balancing this by also dividing the top of the fraction by $\sqrt{2}$. This gives

$$\int_{0.0}^{0.5} \frac{\frac{1}{\sqrt{2}} dx}{\sqrt{\frac{1-2x^2}{2}}} = \frac{1}{\sqrt{2}} \int_{0.0}^{0.5} \frac{dx}{\sqrt{\frac{1}{2}-x^2}}$$

since the constant term $\dfrac{1}{\sqrt{2}}$ can be taken outside the integration.

Comparing the function to be integrated with the general form now shows that $a^2 = \dfrac{1}{2}$ so that $a = \sqrt{\dfrac{1}{2}} = \dfrac{1}{\sqrt{2}}$. Applying the result for the standard integral shows that

$$\int_{0.0}^{0.5} \frac{dx}{\sqrt{1-2x^2}} = \sqrt{\frac{1}{2}}\left[\sin^{-1}\left(\frac{x}{\frac{1}{\sqrt{2}}}\right)\right]_{0.0}^{0.5}$$

$$= \sqrt{\frac{1}{2}}\left[\sin^{-1}(x\sqrt{2})\right]_{0.0}^{0.5}$$

$$= \sqrt{\frac{1}{2}}\left[\sin^{-1}0.5\sqrt{2} - \sin^{-1}0.0\sqrt{2}\right]$$

$$= 0.707\,1\left[\sin^{-1}0.707\,1 - \sin^{-1}0.0\right]$$

$$= 0.707\,1\left[0.785\,4 - 0.000\,0\right]$$

$$= 0.555\,4$$

Such calculations normally require the result of the inverse trigonometric function to be expressed in radians, requiring the appropriate mode to be set on a calculator.

Worked Example 7.9

The expectation value of the position of the particle within the box, <x>, is given by the equation

$$< x > = \frac{2}{a} \int_0^a x \sin^2 \left(\frac{n\pi x}{a} \right) dx$$

Determine the value of <x> given the standard integral

$$\int x \sin^2 kx \, dx = \frac{x^2}{4} - \frac{x}{4k} \sin 2kx - \frac{1}{8k^2} \cos 2kx + C$$

CHEMICAL BACKGROUND

The expectation value of a quantity is its average experimental value. It clearly gives us a link between the sometimes abstract nature of quantum mechanics and the real world. In some cases, expectation values reinforce our classical view of systems, while in others they provide surprising results which involve us in further thought.

Solution to Worked Example 7.9

We begin by recasting the standard integral given into the form suggested by the particular problem. In this case, the variable x is multiplied by the constant $\frac{n\pi}{a}$; we thus replace k in the standard integral so that it becomes

$$\int x \sin^2 \left(\frac{n\pi x}{a} \right) dx = \frac{x^2}{4} - \frac{x}{4 \left(\frac{n\pi}{a} \right)} \sin 2 \left(\frac{n\pi}{a} \right) x$$

$$- \frac{1}{8 \left(\frac{n\pi}{a} \right)^2} \cos 2 \left(\frac{n\pi}{a} \right) x + C$$

which can be rearranged to give

$$\int x \sin^2 \left(\frac{n\pi}{a} \right) x dx = \frac{x^2}{4} - \frac{ax}{4n\pi} \sin 2 \left(\frac{n\pi}{a} \right) x - \frac{a^2}{8n^2\pi^2} \cos 2 \left(\frac{n\pi}{a} \right) x + C$$

The definite integral we require is then

$$\int_0^a x\sin^2\left(\frac{n\pi}{a}\right)xdx = \left[\frac{x^2}{4} - \frac{ax}{4n\pi}\sin 2\left(\frac{n\pi}{a}\right)x - \frac{a^2}{8n^2\pi^2}\cos 2\left(\frac{n\pi}{a}\right)x\right]_0^a$$

$$= \left[\frac{a^2}{4} - \frac{a^2}{4n\pi}\sin 2n\pi - \frac{a^2}{8n^2\pi^2}\cos 2n\pi\right] - \left[0 - 0 - \frac{a^2}{8n^2\pi^2}\cos 0\right]$$

Since n can only take integral values, $\sin 2n\pi = 0$ and $\cos 2n\pi = 1$. Therefore

$$\int_0^a x\sin^2\left(\frac{n\pi}{a}\right)xdx = \left[\frac{a^2}{4} - 0 - \frac{a^2}{8n^2\pi^2}\times 1\right] - \left[-\frac{a^2}{8n^2\pi^2}\times 1\right]$$

$$= \frac{a^2}{4} - \frac{a^2}{8n^2\pi^2} + \frac{a^2}{8n^2\pi^2}$$

$$= \frac{a^2}{4}$$

Consequently

$$<x> = \frac{2}{a}\times\frac{a^2}{4}$$

$$= \frac{a}{2}$$

You will probably not be surprised by the result that the expectation value for the position of the particle is in the centre of the box.

7.6 THE FREE PARTICLE

Although a free particle which is not subject to any forces is not a particularly interesting system from a chemical point of view, it does allow us to work through a quantum mechanical treatment at a relatively simple level. A knowledge of the potential energy of the system allows the Schrödinger equation to be written and subsequently solved to give the general wavefunction which contains all the information needed to describe the behaviour of the system. We then need to apply the boundary conditions and normalize the function before it can be used to predict properties.

7.6.1 THE COMPLEX CONJUGATE

We saw in Section 6.6.1 on page 224 that a complex number z was defined in terms of real numbers a and b as

$$z = a + ib$$

where $i^2 = -1$. The complex conjugate of z, known as z^*, is defined as

$$z^* = a - ib$$

So, to obtain the conjugate of a complex number, we change the sign of the imaginary part. If $z = 2 + 3i$, we would then have $z^* = 2 - 3i$. Similarly, if $z = 2 - 4i$, we would have $z^* = 2 + 4i$. z and z^* are called a **complex conjugate pair;** each one is the complex conjugate of the other.

Notice that since

$$e^{ikx} = \cos kx + i \sin kx$$

and

$$e^{-ikx} = \cos kx - i \sin kx$$

e^{ikx} and e^{-ikx} are a complex conjugate pair.

7.6.2 THE MODULUS OF A COMPLEX NUMBER

We saw in the previous chapter that the modulus of a vector was its length, without regard to direction. Complex numbers can be represented in a similar fashion to vectors by the use of what is known as an Argand diagram. This plots the real part a of a complex number in the x-direction and the imaginary part b in the y-direction. Such a plot for the complex number $1 + 4i$ is shown in Figure 7.10.

The modulus of the complex number z is then given as

$$|z| = \sqrt{a^2 + b^2}$$

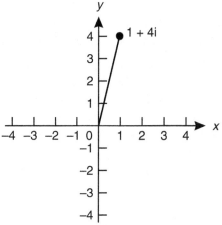

FIGURE 7.10   The Argand diagram for $1 + 4i$.

This can be compared with the expression for calculating the modulus of a vector which we met in Section 6.5.3. If we now consider the product of z^*, the complex conjugate of z, with z, we get

$$
\begin{aligned}
z^*z &= (a - ib)(a + ib) \\
&= a(a + ib) - ib(a + ib) \\
&= a^2 + aib - iba - i^2b^2
\end{aligned}
$$

Since $aib = iba$ and $i^2 = -1$ we obtain

$$z^*z = a^2 + b^2$$

and taking the square root of both sides of this equation gives

$$\sqrt{z^*z} = \sqrt{a^2 + b^2} = |z|$$

so that the modulus of a complex number can be obtained by taking the square root of the product of the complex number and its complex conjugate.

Worked Example 7.10

One wavefunction Ψ which satisfies the Schrödinger equation for a free particle which moves only in the x-direction is given by the equation

$$\Psi = A\exp\left(-i\sqrt{8\pi^2 mE}\,\frac{x}{h}\right)$$

where A is a constant, h is Planck's constant, m is the mass of the particle and E its energy when at position x. The probability of finding the particle at a particular position is given by the function $\Psi^*\Psi$. Determine the equation which describes this function.

CHEMICAL BACKGROUND

An example of a free particle in chemistry is an ionized electron, which may have any energy value. The Balmer series in the spectrum of hydrogen converges when such electrons fall to the energy level with principal quantum number 2.

Since the final value for the probability does not depend on x, the free particle has an equal probability of being found anywhere in the x-direction, and it is said to be nonlocalized.

Solution to Worked Example 7.10

From the expression for the wavefunction

$$\Psi = A\exp\left(-i\sqrt{8\pi^2 mE}\,\frac{x}{h}\right)$$

we can immediately write down an expression for the conjugate wavefunction

$$\Psi^* = A^* \exp\left(i\sqrt{8\pi^2 mE}\, \frac{x}{h} \right)$$

where A^* is the complex conjugate of A and $-i$ has been replaced by i. We are now able to form the product of these two functions:

$$\Psi^*\Psi = A^* \exp\left(i\sqrt{8\pi^2 mE}\, \frac{x}{h} \right) A \exp\left(-i\sqrt{8\pi^2 mE}\, \frac{x}{h} \right)$$

and since

$$e^{-x} = \frac{1}{e^x}$$

from Section 1.6.5 on page 19, this can be rewritten as

$$\Psi^*\Psi = A^* A \frac{\exp\left(i\sqrt{8\pi^2 mE}\, \dfrac{x}{h} \right)}{\exp\left(i\sqrt{8\pi^2 mE}\, \dfrac{x}{h} \right)}$$

$$= A^* A$$

$$= |A|^2$$

7.7 THE HYDROGEN ATOM WAVEFUNCTION

In its ground state, the single electron in a hydrogen atom occupies the 1s shell; this could be promoted to the 2s or 2p level in the excited state. The numbers 1 and 2 are known as the principal quantum number.

Wavefunctions can be defined for the electron in a hydrogen atom in both its ground state and in these higher energy states. Their form is found by solving the appropriate Schrödinger equation, and details of this procedure can be found in physical chemistry textbooks. The wavefunctions are normally expressed as functions of three variables, which are the distance r of the electron from the nucleus, the zenith angle θ and the azimuthal angle ϕ, as shown in Figure 7.11. These quantities are known as spherical polar coordinates and define the position of a particle in space, and are more useful than the more familiar cartesian coordinates x, y and z when we are dealing with systems such as this where the symmetry is spherical.

7.7.1 DIFFERENTIATION OF A PRODUCT

We saw in Section 3.5.1 on page 71 how to differentiate simple functions. We may also come across *products* of functions which need to be

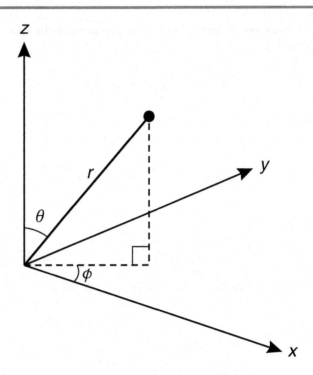

FIGURE 7.11 Definition of the spherical coordinates r, θ and ϕ.

differentiated, such as $x^2 e^x$. In this case, we need to be able to apply the standard rule for the differentiation of a product.

If our function $f(x)$ can be expressed as the product of two functions $u(x)$ and $v(x)$, such that

$$f(x) = u(x)v(x)$$

then

$$\frac{df(x)}{dx} = v(x)\frac{du(x)}{dx} + u(x)\frac{dv(x)}{dx}$$

This is perhaps more easily remembered in words as 'differentiate the first and multiply by the second, differentiate the second and multiply by the first'.

In the example above of $f(x) = x^2 e^x$, we set

$$u(x) = x^2, \; v(x) = e^x$$

and

$$\frac{du(x)}{dx} = 2x, \frac{dv(x)}{dx} = e^x$$

so

$$\frac{d(x^2 e^x)}{dx} = e^x \times 2x + x^2 \times e^x$$

$$= 2xe^x + x^2 e^x$$

7.7.2 INTEGRATION BY PARTS

The equation expressing the rule for differentiating a product can also be of use to us when integrating certain more complicated functions. Writing this with the omission of the function notation for clarity as

$$\frac{d(uv)}{dx} = v\frac{du}{dx} + u\frac{dv}{dx}$$

allows us to rearrange to

$$u\frac{dv}{dx} = \frac{d(uv)}{dx} - v\frac{du}{dx}$$

by subtracting the term $v\dfrac{du}{dx}$ from both sides.

We can now integrate this expression to give the equation

$$\int u\frac{dv}{dx}dx = \int \frac{d(uv)}{dx}dx - \int v\frac{du}{dx}dx$$

Since integration is the reverse of differentiation, integrating a derivative gives back the original function and so we have

$$\int \frac{d(uv)}{dx}dx = uv$$

which gives us

$$\int u\frac{dv}{dx}dx = uv - \int v\frac{du}{dx}dx$$

It is useful to remember that u is a function that we will need to differentiate, and $\dfrac{dv}{dx}$ is a function we will need to integrate. This equation is used to perform the integration process known as **integration by parts.**

If we consider the calculation of the integral

$$\int x\ln x\, dx$$

we can integrate and differentiate the term x, but with $\ln x$ it is much easier to differentiate. So, if we make the substitutions

$$u = \ln x \text{ and } \frac{dv}{dx} = x$$

we can differentiate u and integrate $\dfrac{dv}{dx}$, using the rules in Sections 5.6.2 and 5.4.1 respectively, to give

$$\frac{du}{dx} = \frac{1}{x} \text{ and } v = \frac{x^2}{2}$$

ignoring, for the time being, the constant of integration. We can now substitute into the general equation

$$\int u \frac{dv}{dx} dx = uv - \int v \frac{du}{dx} dx$$

to give

$$\int x \ln x \, dx = \frac{x^2}{2} \ln x - \int \frac{x^2}{2} \frac{1}{x} dx$$

$$= \frac{x^2}{2} \ln x - \int \frac{x}{2} dx$$

$$= \frac{x^2}{2} \ln x - \frac{x^2}{4} + C$$

where C is the constant of integration, normally inserted at the end of the calculation for simplicity. Be aware that it is easy to forget it!

7.7.3 CALCULUS OF THE EXPONENTIAL FUNCTION

The exponential function e^x is the only function which does not change upon differentiation and integration. We now need to be aware of the slightly more general result for the function e^{ax} where a is a constant. Examples are e^{3x} or e^{-2x} The rules we need are

$$\frac{d(e^{ax})}{dx} = ae^{ax} \text{ and } \int e^{ax} dx = \frac{e^{ax}}{a} + C$$

Some examples of these are

$$\frac{d(e^{-x})}{dx} = -e^{-x}$$

$$\frac{d(e^{4x})}{dx} = 4e^{4x}$$

$$\int e^{-3x} dx = \frac{e^{-3x}}{-3} + C$$

Worked Example 7.11

During the normalization of the wavefunction representing the lowest electronic state of the hydrogen atom, the integral

$$\int_0^\infty r^2 \exp\left(-\frac{2r}{a_o}\right) dr$$

needs to be evaluated. Use integration by parts to evaluate this integral.

CHEMICAL BACKGROUND

The process of normalization is used to ensure that the probability of finding the specified particle within a given volume is correct. Since the electron must be found somewhere within the hydrogen atom, the probability of finding it is 1, and so the probability function derived from the wavefunction must integrate over all space to give this value. It can be adjusted by the inclusion of a normalization factor, which is usually given the symbol N. Note that we integrate from $r = 0$ to infinity ∞ as we do not know at what distance the probability falls to zero.

Solution to Worked Example 7.11

The function to be integrated is shown in Figure 7.12. We are obviously going to use the equation given above for performing integration by parts. This problem is given in terms of the variable r, so the equation we need to use is actually

$$\int u \frac{dv}{dr} dr = uv - \int v \frac{du}{dr} dr$$

where x in our general equation has been replaced by r which is the variable specific to this example.

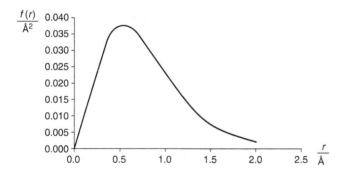

FIGURE 7.12 Graph of the function $f(r)$ integrated into Worked Example 7.11.

We now need to assign the two functions $u(r)$ and $v(r)$ to the two functions r^2 and $\exp\left(-\dfrac{2r}{a_o}\right)$. If we set $u(r) = r^2$, then we obtain, by differentiation

$$\frac{du}{dr} = 2r$$

whereas if we set $\dfrac{dv}{dr} = r^2$ we obtain, by integration

$$v(r) = \frac{1}{3}r^3$$

The first of these seems to be leading towards a simpler calculation, but we also need to consider what happens to the second function.

If we set $u(r) = \exp\left(-\dfrac{2r}{a_o}\right)$ then, by differentiating, we obtain

$$\frac{du}{dr} = -\frac{2}{a_o}\exp\left(-\frac{2r}{a_o}\right)$$

whereas the substitution $\dfrac{dv}{dr} = \exp\left(-\dfrac{2r}{a_o}\right)$ gives, on integration

$$v(r) = \frac{\exp\left(-\dfrac{2r}{a_o}\right)}{-\dfrac{2}{a_o}}$$

$$= -\frac{a_o}{2}\exp\left(-\frac{2r}{a_o}\right)$$

This suggests that setting

$$u(r) = r^2 \text{ and } \frac{dv}{dr} = \exp\left(-\frac{2r}{a_o}\right)$$

so that

$$\frac{du}{dr} = 2r \text{ and } v = -\frac{a_o}{2}\exp\left(-\frac{2r}{a_o}\right)$$

leads to a simpler calculation.

Making these substitutions in our general equation

$$\int u\frac{dv}{dr}dr = uv - \int v\frac{du}{dr}dr$$

gives

$$\int r^2 \exp\left(-\frac{2r}{a_o}\right) = r^2 \left(-\frac{a_o}{2}\exp\left(-\frac{2r}{a_o}\right)\right) - \int \left(-\frac{a_o}{2}\exp\left(-\frac{2r}{a_o}\right)\right) 2r\,dr$$

$$= -\frac{a_o}{2}r^2 \exp\left(-\frac{2r}{a_o}\right) + a_o \int r \exp\left(-\frac{2r}{a_o}\right) dr$$

At this point in a problem involving the use of integration by parts, we would now evaluate the integral in the second term on the right of the equation which is

$$\int r \exp\left(-\frac{2r}{a_o}\right) dr$$

In this case, however, it is not at all obvious how this can be done. A closer inspection reveals that this integral can also be evaluated using integration by parts. If we set

$$u = r \text{ and } \frac{dv}{dr} = \exp\left(-\frac{2r}{a_o}\right)$$

We have

$$\frac{du}{dr} = 1 \text{ and } v(r) = \frac{\exp\left(-\frac{2r}{a_o}\right)}{-\frac{2}{a_o}} = -\frac{a_o}{2}\exp\left(-\frac{2r}{a_o}\right)$$

Substituting these values into

$$\int u \frac{dv}{dr} dr = uv - \int v \frac{du}{dr} dr$$

gives us

$$\int r \exp\left(-\frac{2r}{a_o}\right) dr = -\frac{a_o}{2}r\exp\left(-\frac{2r}{a_o}\right) - \int \left(-\frac{a_o}{2}\exp\left(-\frac{2r}{a_o}\right)\right) \times 1\,dr$$

$$= -\frac{a_o}{2}r\exp\left(-\frac{2r}{a_o}\right) + \frac{a_o}{2}\int \exp\left(-\frac{2r}{a_o}\right) dr$$

$$= -\frac{a_o}{2}r\exp\left(-\frac{2r}{a_o}\right) + \frac{a_o}{2}\frac{\exp\left(-\frac{2r}{a_o}\right)}{-\frac{2}{a_o}}$$

$$= -\frac{a_o}{2}r\exp\left(-\frac{2r}{a_o}\right) - \frac{a_o^2}{4}\exp\left(-\frac{2r}{a_o}\right)$$

Since we have been able to evaluate this integral completely, the expression can now be substituted into our equation for the original integral to give

$$\int r^2 \exp\left(-\frac{2r}{a_o}\right) dr = -\frac{a_o}{2} r^2 \exp\left(-\frac{2r}{a_o}\right)$$

$$+ a_o \left(-\frac{a_o}{2} r \exp\left(-\frac{2r}{a_o}\right) - \frac{a_o^2}{4} \exp\left(-\frac{2r}{a_o}\right)\right)$$

$$= -\frac{a_o}{2} r^2 \exp\left(-\frac{2r}{a_o}\right) - \frac{a_o^2}{2} r \exp\left(-\frac{2r}{a_o}\right)$$

$$- \frac{a_o^3}{4} \exp\left(-\frac{2r}{a_o}\right)$$

$$= \exp\left(-\frac{2r}{a_o}\right) \left(-\frac{a_o r^2}{2} - \frac{a_o^2 r}{2} - \frac{a_o^3}{4}\right)$$

We have not yet included the limits on our integration sign. It is actually easier to include them only at the end, so from our previous work we have

$$\int_0^\infty r^2 \exp\left(-\frac{2r}{a_o}\right) dr = \left[\exp\left(-\frac{2r}{a_o}\right)\left(-\frac{a_o r^2}{2} - \frac{a_o^2 r}{2} - \frac{a_o^3}{4}\right)\right]_0^\infty$$

Putting in an upper limit of $r = \infty$ first, and realizing that

$$\exp\left(-\frac{2r}{a_o}\right) = \frac{1}{\exp\left(\dfrac{2r}{a_o}\right)}$$

since $e^{-x} = \dfrac{1}{x}$ (Section 1.6.5 on page 19) shows that the whole of this term will tend to zero for the upper limit, since we have the reciprocal of a very large number. Multiplying by zero, gives a value of zero for the whole expression as r tends to infinity ∞.

On the other hand, when $r = 0$

$$\exp\left(-\frac{2r}{a_o}\right) = e^0 = 1$$

since any number raised to the power zero is one. The first two terms in brackets are zero when $r = 0$, and so we are left with

$$\int_0^\infty r^2 \exp\left(-\frac{2r}{a_o}\right) dr = 0 - \left(-\frac{a_o^3}{4}\right)$$

$$= \frac{a_o^3}{4}$$

7.7.4 MULTIPLE INTEGRATION

In Section 3.7.6 on page 99, we met the idea of a partial derivative, where a function $f(x, y)$ of two variables can be differentiated with respect to either variable while keeping the other constant. Differentiating $f(x, y)$ with respect to x while keeping y constant gives the derivative denoted by

$$\left(\frac{\partial f}{\partial x}\right)_y$$

while differentiating $f(x, y)$ with respect to y while keeping x constant leads to the derivative noted by

$$\left(\frac{\partial f}{\partial y}\right)_x$$

When we consider the integration of this function, we can determine

$$\int_a^b f(x, y) dx$$

where $f(x, y)$ is integrated with respect to x between $x = a$ and $x = b$ while keeping y constant, or

$$\int_a^b f(x, y) dy$$

where $f(x, y)$ is integrated with respect to y between $y = a$ and $y = b$ while keeping x constant.

Alternatively, we can also determine the double integral of $f(x, y)$ with respect to both x and y. This would be denoted as

$$\int_c^d \int_a^b f(x, y) dx\, dy$$

By convention, this is read from the outside in. Thus the constants c and d are the lower and upper limits respectively when integrating with respect to y, while a and b are the respective lower and upper limits when integrating with respect to x. The process involved may be clearer if rewritten using brackets to give

$$\int_c^d \left(\int_a^b f(x,y)\,dx \right) dy$$

Using the convention that we evaluate quantities in brackets first, we begin by integrating $f(x, y)$ with respect to x between limits a and b, then integrate the result of this with respect to y between limits c and d.

Applying this to a double integral such as

$$\int_0^1 \int_2^3 (3xy + 2x)\,dx\,dy$$

we begin by integrating $3xy + 2x$ with respect to x to give

$$\int_2^3 (3xy + 2x)\,dx = \left[\frac{3x^2 y}{2} + \frac{2x^2}{2} \right]_2^3$$

$$= \left[\frac{3x^2 y}{2} + x^2 \right]_2^3$$

$$= \left[\frac{3 \times 3^2 y}{2} + 3^2 \right] - \left[\frac{3 \times 2^2 y}{2} + 2^2 \right]$$

$$= \left[\frac{3 \times 9y}{2} + 9 \right] - \left[\frac{3 \times 4y}{2} + 4 \right]$$

$$= \left[\frac{27y}{2} + 9 \right] - \left[\frac{12y}{2} + 4 \right]$$

$$= \frac{15y}{2} + 5$$

This can now be integrated with respect to y so we have

$$\int_0^1 \int_2^3 (3xy + 2x)\,dx\,dy = \int_0^1 \left(\frac{15y}{2} + 5 \right) dy$$

$$= \left[\frac{15y^2}{2} + 5y \right]_0^1$$

$$= \left[\frac{15 \times 1^2}{2} + 5 \times 1 \right] - \left[\frac{15 \times 0^2}{2} + 5 \times 0 \right]$$

$$= \left[\frac{15}{2} + 5 \right] - [0]$$

$$= \frac{25}{2}$$

$$= 12.5$$

It is also possible to evaluate triple integrals in a similar fashion.

7.7.5 CALCULUS OF THE TRIGONOMETRIC FUNCTIONS

The trigonometric functions and their inverses were introduced in Section 6.4 on page 207. In particular, Figure 6.6 showed the graphs of the sine and cosine functions. Since we can plot these graphs it follows that we can determine the gradient of the function at any point. This, as we saw in Section 3.5.1 on page 71 is equivalent to determining the derivative. Consequently, we can discuss the calculus of the trigonometric functions.

In this section, we will merely note the results of differentiating and integrating the general sine and cosine functions. In their most general forms these are:

$$\frac{d}{dx}(\sin(ax+b)) = a\cos(ax+b)$$

$$\frac{d}{dx}(\cos(ax+b)) = -a\sin(ax+b)$$

$$\int \sin(ax+b)\,dx = -\frac{1}{a}\cos(ax+b) + C$$

$$\int \cos(ax+b) = \frac{1}{a}\sin(ax+b) + C$$

where C is the constant of integration.

As an example of differentiation, consider

$$\frac{d}{dx}(\cos(3x-2))$$

Comparing with the general form of the function $\cos(ax + b)$ we see that we have $a = 3$ and $b = -2$ and so

$$\frac{d}{dx}(\cos(3x-2)) = -3\sin(3x-2)$$

noting in particular the negative sign. If we now consider

$$\int_0^\pi \cos 5x\,dx$$

and compare with the general function $\cos(ax + b)$ we see that $a = 5$ and $b = 0$ so that

$$\int \cos 5x\,dx = \frac{1}{5}\sin 5x$$

However, we are determining the definite integral, so we need to consider the lower and upper limits which are 0 and π respectively. Unless specified, such limits on trigonometric functions are taken as being in radians. We then have

$$\int_0^\pi \cos 5x\, dx = \left[\frac{1}{5}\sin 5x\right]_0^\pi$$

$$= \frac{1}{5}\sin 5\pi - \frac{1}{5}\sin 0$$

$$= 0 - 0$$

$$= 0$$

since the value of the sine of any multiple of π rad (180°) is zero.

Worked Example 7.12

Determine the integral

$$\int_0^{2\pi}\int_0^\pi\int_0^\infty e^{-\frac{2r}{a_o}} r^2 \sin\theta\, dr\, d\theta\, d\phi$$

which is required in order to evaluate the normalization constant for the hydrogen atom wavefunction.

CHEMICAL BACKGROUND

Notice that this equation is formulated in terms of the spherical polar coordinates introduced in Section 7.7 and shown in Figure 7.11. As we have seen earlier in this section, the normalization constant is introduced in order to ensure that the probability of finding an electron anywhere in space is exactly 1. This is expressed mathematically by the equation

$$\int_{-\infty}^{\infty} \Psi^*\Psi = 1$$

where Ψ is the wavefunction and Ψ* its complex conjugate (as defined in Section 7.6.1), the integration limits being taken from minus infinity to infinity to represent the whole of space.

Solution to Worked Example 7.12

When we integrate this expression with respect to r the only part of the function involved is $e^{-\frac{2r}{a_o}} r^2$ which can be rewritten as $r^2 e^{-\frac{2r}{a_o}}$. Similarly, integration with respect to θ involves only sin θ. We can therefore rewrite the expression to be integrated as

$$\int_0^{2\pi}\left(\int_0^\pi \sin\theta\left(\int_0^\infty r^2 e^{-\frac{2r}{a_o}}\, dr\right)d\theta\right)d\phi$$

Using brackets to make the order of integration clearer. Working from the outside in, the integral of $d\phi$ is ϕ, so we have

$$\int_0^{2\pi}\left(\int_0^{\pi}\sin\theta\left(\int_0^{\infty}r^2e^{-\frac{2r}{a_o}}dr\right)d\theta\right)d\phi=\left(\int_0^{\pi}\sin\theta\left(\int_0^{\infty}r^2e^{-\frac{2r}{a_o}}dr\right)d\theta\right)[\phi]_0^{2\pi}$$

$$=2\pi\left(\int_0^{\pi}\sin\theta\left(\int_0^{\infty}r^2e^{-\frac{2r}{a_o}}dr\right)d\theta\right)$$

once the integration limits for ϕ are inserted.

As we saw earlier in this section

$$\int\sin(ax+b)dx=-\frac{1}{a}\cos(ax+b)+C$$

so

$$\int\sin x\,dx=-\cos x+C$$

with $a=b=1$. Therefore

$$\int_0^{\pi}\sin\theta\,d\theta=\left[-\cos\theta\right]_0^{\pi}$$

$$=\left[-\cos\pi\right]-\left[-\cos 0\right]]$$

$$=[-(-1)]-[-1]$$

$$=1+1$$

$$=2$$

remembering that θ is in this case expressed in radians, and that the two adjacent negative signs combine to give a positive. Our integral is now

$$2\pi\times 2\left(\int_0^{\infty}r^2e^{-\frac{2r}{a_o}}dr\right)$$

The result of Worked example 7.11 now allows us to write

$$\int_0^{2\pi}\left(\int_0^{\pi}\sin\theta\left(\int_0^{\infty}r^2e^{-\frac{2r}{a_o}}dr\right)d\theta\right)d\phi=2\pi\times 2\times\frac{a_o^3}{4}$$

$$=\pi a_o^3$$

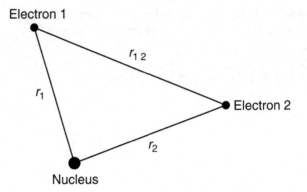

FIGURE 7.13   Interactions between particles in the helium atom.

7.8 THE HELIUM ATOM

The quantum mechanics of the helium atom are rather more complicated than those of the hydrogen atom because we now have to account for the interactions between electrons, as well as that between an electron and the nucleus. These interactions are illustrated in Figure 7.13. In practice, we build a trial wavefunction based on the information we have already obtained for the 1s orbital of hydrogen.

7.8.1 STATIONARY POINTS

The method for determining the stationary points of a function was presented in Section 4.8.2 on page 139. The stationary points of a function $f(x)$ occur when

$$\frac{df(x)}{dx} = 0$$

and

$$\frac{d^2 f(x)}{dx^2} < 0 \text{ for a maximum}$$

while

$$\frac{d^2 f(x)}{dx^2} > 0 \text{ for a minimum.}$$

Worked example 7.13

The energy E of the helium atom can be expressed as

$$E = \left(\frac{-2\pi^2 m_e e^4}{h^2} \right) \left(-2(Z')^2 + \frac{27}{4} Z' \right)$$

where m_e is the mass of an electron, e is the electronic charge and h is Planck's constant. Z' is the effective nuclear charge on the helium atom and is treated as an adjustable parameter. Find the value of Z' which makes the value of E a minimum.

CHEMICAL BACKGROUND

This approach leads to a value for the energy of -77.5 eV which compares with an experimental value of -79.0 eV. A calculated value of -78.7 eV can be obtained by including a term for electron correlation in the 1s orbital trial wavefunction. The function for E used here is shown in Figure 7.14.

The method being used here is actually an example of the Variation Theorem, which finds a number of applications in quantum mechanics. It involves specifying a trial wavefunction and then optimizing a specified adjustable parameter.

Solution to Worked Example 7.13

We will obtain a minimum value of E when

$$\frac{dE}{dZ'} = 0 \text{ and } \frac{d^2 E}{dZ'^2} > 0$$

The term in the first bracket of the expression for E is a constant and so this remains as a multiplier throughout the calculation. We know that

$$\frac{d(x^2)}{dx} = 2x \text{ and } \frac{d(x)}{dx} = 1$$

so that

$$\frac{d}{dZ'}\left(-2(Z')^2 + \frac{27}{4}Z'\right) = -4Z' + \frac{27}{4}$$

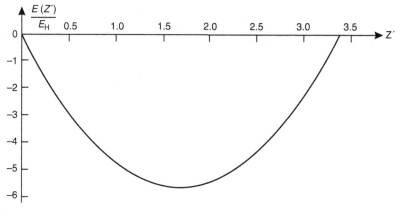

FIGURE 7.14 Graph of the function representing the energy of the helium atom. $E_H = 27.2$ eV.

This term will be zero when

$$-4Z' + \frac{27}{4} = 0$$

Adding $4Z'$ to both sides and dividing both sides by 4, isolates Z' to give

$$Z' = \frac{27}{16}$$

The full expression for the first derivative of E is

$$\frac{dE}{dZ'} = \left(\frac{-2\pi^2 m_e e^4}{h^2} \right)\left(-4Z' + \frac{27}{4} \right)$$

so the second derivative will be

$$\frac{d^2 E}{dZ'^2} = -4\left(\frac{-2\pi^2 m_e e^4}{h^2} \right) = \frac{8\pi^2 m_e e^4}{h^2}$$

since $\frac{d}{dx}(x) = 1$. This is a positive value, and the second derivative being positive confirms that this stationary point is a minimum.

7.9 HÜCKEL THEORY

As molecules become larger it is harder to treat them using quantum mechanics. Hückel theory applies a fairly drastic set of approximations to conjugated systems by assuming that σ-bonds can effectively be ignored and only treating the π-bonds. It distinguishes between the following types of integrals:

1. Overlap integrals which are set to zero.
2. Resonance integrals, which are set to zero between non-neighbouring atoms and to a constant β between neighbouring atoms.
3. Coulomb integrals which are set to a constant α.

These interactions can be described in terms of a matrix for each system.

7.9.1 DETERMINANTS

The determinant of a 2×2 matrix

$$\begin{pmatrix} a & b \\ c & d \end{pmatrix}$$

is denoted as

$$\begin{vmatrix} a & b \\ c & d \end{vmatrix}$$

FIGURE 7.15 Diagonals used to calculate a 2 × 2 determinant.

and is equal to *ad-bc*, i.e. the leading diagonal (top left to bottom right) minus the trailing diagonal (top right to bottom left) as shown in Figure 7.15. For example

$$\begin{vmatrix} 3 & 4 \\ 2 & 6 \end{vmatrix} = (3\times 6)-(4\times 2)=18-8=10$$

To obtain the determinant of a 3 × 3 matrix we have

$$\begin{vmatrix} a & b & c \\ d & e & f \\ g & h & i \end{vmatrix} = a\begin{vmatrix} e & f \\ h & i \end{vmatrix} - b\begin{vmatrix} d & f \\ g & i \end{vmatrix} + c\begin{vmatrix} d & e \\ g & h \end{vmatrix}$$

FIGURE 7.16 Quantities used to calculate a 3 × 3 determinant.

You will see that we have worked along the top row alternating the sign so that we have three terms in a, −b and c. For each of these, we have then multiplied by the 2 × 2 determinant obtained by eliminating the row and column containing the first number as shown in Figure 7.16. So, in the first term, we have eliminated the first row and first column containing *a*. Expanding each of these 2 × 2 determinants then gives

$$\begin{vmatrix} a & b & c \\ d & e & f \\ g & h & i \end{vmatrix} = a(ei-fh)-b(di-fg)+c(dh-eg)$$

For example

$$\begin{vmatrix} 3 & 2 & 4 \\ -2 & 1 & 5 \\ -3 & 2 & 3 \end{vmatrix} = 3\begin{vmatrix} 1 & 5 \\ 2 & 3 \end{vmatrix} - 2\begin{vmatrix} -2 & 5 \\ -3 & 3 \end{vmatrix} + 4\begin{vmatrix} -2 & 1 \\ -3 & 2 \end{vmatrix}$$

$$= 3(1\times 3-5\times 2)-2\big((-2)\times 3-5\times(-3)\big)+4\big((-2)\times 2-1\times(-3)\big)$$

$$= 3(3-10)-2(-6+15)+4(-4+3)$$

$$= 3\times(-7)-2\times 9+4\times(-1)$$

$$= -21-18-4 = -43$$

A similar technique is used for evaluating higher-order determinants, but the calculations do become more complicated. It is easier when a significant number of the matrix elements are zero.

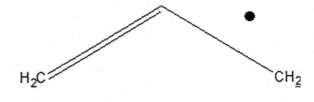

FIGURE 7.17   The allyl radical.

Worked Example 7.14

In the Hückel treatment of the allyl radical (Figure 7.17) the following equation needs to be solved:

$$\begin{vmatrix} \alpha - E & \beta & 0 \\ \beta & \alpha - E & \beta \\ 0 & \beta & \alpha - E \end{vmatrix} = 0$$

By evaluating this determinant and setting equal to zero obtain the values of E.

CHEMICAL BACKGROUND

While this may seem like an unusual species to study, it is the three-carbon atom structure that is amenable to treatment by Hückel theory.

Solution to Worked Example 7.14

We work along the top row of the matrix as shown in the earlier example. This gives

$$(\alpha - E)\begin{vmatrix} \alpha - E & \beta \\ \beta & \alpha - E \end{vmatrix} - \beta \begin{vmatrix} \beta & \beta \\ 0 & \alpha - E \end{vmatrix} - 0 \begin{vmatrix} \beta & \alpha - E \\ 0 & \beta \end{vmatrix} = 0$$

We are able to ignore the third term which will be zero, so expanding the remaining 2 × 2 determinants leads to

$$(\alpha - E)\left\{(\alpha - E)^2 - \beta^2\right\} - \beta\left\{\beta(\alpha - E) - \beta \times 0\right\} = 0$$

and so

$$(\alpha - E)^3 - \beta^2(\alpha - E) - \beta^2(\alpha - E) = 0$$

or

$$(\alpha - E)^3 - 2\beta^2(\alpha - E) = 0$$

If we divide this equation by β^3 we obtain

$$\left(\frac{\alpha - E}{\beta}\right)^3 - 2\left(\frac{\alpha - E}{\beta}\right) = 0$$

If we now set

$$x = \frac{\alpha - E}{\beta}$$

This becomes

$$x^3 - 2x = 0$$

which can also be written as

$$x(x^2 - 2) = 0$$

The solutions to this equation are $x = 0$ and $x^2 = 2$. Because the square root of a number can be positive or negative, possible solutions are:

$$x = 0 \text{ so } \frac{\alpha - E}{\beta} = 0, \alpha = E.$$

$$x = \sqrt{2} \text{ so } \frac{\alpha - E}{\beta} = \sqrt{2}, \alpha - E = \sqrt{2}\beta, E = \alpha - \sqrt{2}\beta$$

$$x = -\sqrt{2} \text{ so } \frac{\alpha - E}{\beta} = -\sqrt{2}, \alpha - E = -\sqrt{2}\beta, E = \alpha + \sqrt{2}\beta$$

EXERCISES

1. Determine the results of the following computations, expressing your answers to the appropriate number of figures:
   a. $9.453\,287\,712 - 0.446\,76$
   b. $3.162\,623\,19 + 5.042\,084\,282\,87$
   c. $0.622\,844\,4 \times 12.309\,421\,722\,306\,987$
   d. $\dfrac{83.591\,914\,614\,927\,751\,1}{4.706\,508\,322}$

2. From the following data deduce the relationship between the variables x and y:

x	1.42	3.62	4.75	6.24	7.08
y	7.12	2.79	2.13	1.62	1.43

3. Two quantities p and q are directly proportional such that

$$\frac{p}{q} = 42.7$$

   a. What is the value of p when $q = 14.3$?
   b. What is the value of q when $p = 0.314$?

4. Determine the stationary points of the function

$$f(x) = 2e^{3x} - 4x$$

5. Determine the location and the nature of the stationary points of the function $f(x) = 2x^2 - 4x + 8$. Hence state the location and nature of the stationary points of $g(x) = 2x^2 - 4x + 2$.

6. Determine the integral

$$\int_{1}^{4}\int_{-1}^{1} (x^2 y + 3xy)\,dx\,dy$$

7. Determine the integral

$$\int_{1}^{2}\int_{0}^{1}\int_{2}^{3} \left(e^x + \frac{1}{z} - \frac{xy}{z^2} \right)\,dx\,dy\,dz$$

8. Give the conjugates of the following complex numbers:
   (a) $4 + 6i$ (b) $-2 - i$ (c) $9 - 4i$ (d) $-3 + 2i$

9. Calculate the product $z_1 z_2$ if

$$z_1 = 3\left(\cos\frac{\pi}{3} - i\sin\frac{\pi}{3} \right)$$

and

$$z_2 = 2\left(\cos\frac{\pi}{4} - i\sin\frac{\pi}{4} \right)$$

10. A sequence is defined by

$$f(n) = 4n^2 - 2n$$

for $n \geq 1$. Give the first five terms of this sequence.

11. Find the next two terms of these sequences:
    a. 4, 7, 10, 13, ..
    b. 5, 10, 17, 26, ..
    c. −4, −3, 1, 10, ..
    d. 1, 7, 17, 31,..

12. Find the general term of the sequence
    3, 6, 12, 24, 48, ...

13. Find the inverses of the functions:
    a. $f(x) = (2x+1)^2$.
    b. $g(x) = (3x - 1)^3$
    c. $h(x) = (x^2 + 3)^2$

14. Find the inverses of the functions:
    a. $f(x) = \ln(x + 2)$
    b. $g(x) = \cos(4x + 1)$
    c. $f(x) = 2e^{ax^2}$.

15. Determine the following derivatives:

    (a) $\dfrac{d}{dx}(x^{0.3})$ (b) $\dfrac{d}{dx}(2x^{-0.6})$ (c) $\dfrac{d}{dy}\left(\dfrac{-4}{y^{1.5}}\right)$ (d) $\dfrac{d}{dt}\left(\dfrac{1}{2t^{\frac{1}{2}}}\right)$

16. Differentiate the functions:
    (a) e^{2x} (b) e^{-3x} (c) $\exp(3x + 2)$ (d) $\exp(2x - 1)$

17. Differentiate these functions:
    a. $f(x) = x^2 \ln x$
    b. $g(x) = x\, e^{2x}$
    c. $h(y) = y^2\, e^{-4y}$
    d. $F(y) = y \sin 3y$

18. Evaluate the following integrals:

    a. $\displaystyle\int e^{4x}\,dx$

    b. $\displaystyle\int_{0}^{3} e^{2x}\,dx$

    c. $\displaystyle\int_{-2}^{-1} e^{-6x}\,dx$

    d. $\displaystyle\int_{-4}^{-3} e^{-3x}\,dx$

19. Use the method of integration by parts to evaluate this definite integral:

$$\int_{0}^{1} xe^{2x}\,dx$$

20. By writing ln x as the product of 1 and ln x, evaluate the integral

$$\int \ln x\,dx$$

    using the method of integration by parts.

21. Calculate the following determinants:

(a) $\begin{vmatrix} 1 & 7 \\ 5 & 2 \end{vmatrix}$ (b) $\begin{vmatrix} 3 & 1 \\ 4 & 1 \end{vmatrix}$ (c) $\begin{vmatrix} 2 & -1 \\ 4 & -3 \end{vmatrix}$ (d) $\begin{vmatrix} -2 & 0 \\ -3 & -4 \end{vmatrix}$

22. Calculate the following determinants:

(a) $\begin{vmatrix} 2 & 0 & 1 \\ 3 & 1 & 2 \\ 1 & 0 & 4 \end{vmatrix}$ (b) $\begin{vmatrix} 1 & 1 & 2 \\ 3 & 0 & 2 \\ 0 & 4 & 1 \end{vmatrix}$ (c) $\begin{vmatrix} 2 & 1 & 3 \\ 1 & 1 & 4 \\ 2 & 1 & 2 \end{vmatrix}$

(d) $\begin{vmatrix} 1 & 0 & 0 & 1 \\ 2 & 1 & 0 & 2 \\ 1 & -1 & 0 & 1 \\ 0 & 0 & 2 & 1 \end{vmatrix}$

PROBLEMS

1. The Bohr radius a_o is defined as

$$a_0 = \frac{4\pi\varepsilon_o\hbar^2}{m_e e^2}$$

where ε_o is the permittivity of a vacuum, m_e is the electron rest mass and e is the elementary charge. The constant $\hbar$ is related to the Planck constant h by the equation

$$\hbar = \frac{h}{2\pi}$$

Determine the value of a_o to an appropriate number of figures, given the values
$\varepsilon_o = 8.854\ 187\ 816 \times 10^{-12}$ F m^{-1}
$h = 6.626\ 070\ 15 \times 10^{-34}$ J s
$m_e = 9.109\ 389\ 7 \times 10^{-31}$ kg
$e = 1.602\ 177\ 33 \times 10^{-19}$ C
You will need to select an appropriate value of π which retains the level of precision implied by this data. Note that the unit Farad, symbol F, is equivalent to the units $C^2 J^{-1}$.

2. A function which can be used to describe the interaction energy $E(r)$ between a pair of molecules a distance r apart is

$$E(r) = \frac{A}{r^9} - \frac{B}{r^6}$$

where A and B are constants. Locate and identify the stationary points of this function.

3. The energy $E(n_x, n_y)$ of a particle in a two-dimensional box is given by the equation

$$E(n_x, n_y) = \frac{h^2}{8m}\left(\frac{n_x^2}{a^2} + \frac{n_y^2}{b^2}\right)$$

where n_x and n_y are quantum numbers in the x and y directions respectively, a and b are the dimensions of the box in the corresponding directions, h is Planck's constant and m is the mass of the particle. Calculate the inverse function arc $E(n)$ in the special case when $n_x = n_y = n$ and $a = b = d$.

4. The following data were obtained for the maximum kinetic energy T_{max} of the electrons emitted from a sodium surface in a photoelectric experiment as a function of frequency v. Determine the relationship between T_{max} and v without using a graph.

T_{max} eV	0.0	0.5	1.0	1.5	2.0	2.5
$10^{-14}v$ Hz	5.47	6.73	7.99	9.25	10.51	11.77

5. The energy E_v of a harmonic oscillator is given by the equation

$$E_v = hv_o\left(v + \tfrac{1}{2}\right)$$

where v is the quantum number which can take values 0, 1, 2, 3.... and v_o is the natural frequency. What are the first five energy levels of the harmonic oscillator?

6. The potential energy V associated with the deformation of a bond angle θ from its natural bond angle θ_o is given by the equation

$$V = \tfrac{1}{2}k_b(\theta - \theta_o)^2$$

where k_b is a constant known as the angle bending force constant. What is the potential energy associated with the deformation of a C-C-H bond from its natural angle by 0.7° if k_b has the value of 33.1 J mol^{-1} degree^{-2}?

7. The wavefunction $\Psi_{1s}(r)$ of the 1s orbital of the hydrogen atom is given by the equation

$$\Psi_{1s}(r) = 2\left(\frac{Z}{a_o}\right)^{\frac{3}{2}} \exp\left(-\frac{Zr}{a_o}\right)$$

where Z is the nuclear charge, a_0 the Bohr radius and r the distance of the electron from the nucleus. Determine the inverse function arc $\Psi_{1s}(r)$.

8. The energy $E(m)$ for the particle on a ring is given in terms of the quantum number m as

$$E(m) = \frac{m^2 h^2}{8\pi^2 I}$$

where I is the moment of inertia and $m = 0, +-1, +-2 ....$ Determine the density of states $\dfrac{dm}{dE}$ for the particle on a ring.

9. The wavefunctions of the particle on a ring are of the form

$$\Psi_m = Ae^{im\phi}$$

where A is a normalization constant. Determine the value of A if the wavefunction satisfies the normalization condition

$$\int_0^{2\pi} \Psi_m^* \Psi_m d\phi = 1$$

where Ψ_m^* is the complex conjugate of Ψ_m.

10. The Variation Theorem is often used to determine the wavefunctions of the helium atom. However, it can also be used with a trial wavefunction for the one-dimensional particle in a box. This involves the integral

$$\int_0^a (ax - x^2)\left(-\frac{h^2}{8\pi^2 m}\right)\frac{d^2}{dx^2}(ax - x^2)dx$$

where a is the length of the box and x is the position of the particle of mass m within it. The constant h is Planck's constant. Perform the required differentiation of the function $ax - x^2$ and then evaluate this integral between the limits shown.

11. The expectation value of x^2 in the one-dimensional particle in a box is given by

$$<x^2> = \frac{2}{a}\int_0^a x^2 \sin^2\left(\frac{n\pi x}{a}\right)dx$$

Determine the value of $<x^2>$ given the standard integral

$$\int x^2 \sin kx \, dx = \frac{x^3}{6} - \left(\frac{x^2}{4k} - \frac{1}{8k^3}\right)\sin 2kx - \frac{x}{4k^2}\cos 2kx + C$$

12. The average distance of an electron from the nucleus in the 1s orbital of the hydrogen atom is given by the expression

$$< r_{1s} >= \frac{1}{\pi a_o^3} \int_0^{2\pi} \int_0^{\infty} \int_0^{\pi} r^3 e^{-\frac{2r}{a_o}} \sin\theta d\theta dr \, d\phi$$

Determine the value of $<r_{1s}>$ given the standard integral

$$\int_0^{\infty} x^3 e^{-kx} dx = \frac{6}{k^4}$$

13. The Hückel determinant for ethene is

$$\begin{vmatrix} \alpha - E & \beta \\ \beta & \alpha - E \end{vmatrix}$$

Solve this equation for E.

14. The Hückel determinant for butadiene (Figure 7.18) is

$$\begin{vmatrix} \alpha - E & \beta & 0 & 0 \\ \beta & \alpha - E & \beta & 0 \\ 0 & b & \alpha - E & \beta \\ 0 & 0 & \beta & \alpha - E \end{vmatrix}$$

Expand this determinant to give an equation in α, β and E which is equal to zero. Solve this equation for E by dividing by β^4 and setting

$$x = \frac{(\alpha - E)^2}{\beta^2}$$

which will give a quadratic equation in x. This will give four values of E.

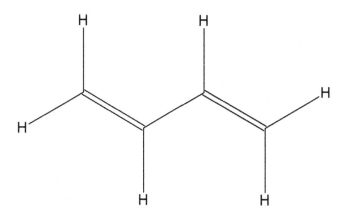

FIGURE 7.18   Structure of butadiene.

Spectroscopy

8.1 INTRODUCTION

The subject of spectroscopy provides us with a description of how chemical particles interact with electromagnetic radiation. Spectroscopic techniques can be classified according to the type of incident radiation (such as ultraviolet or infrared), or according to the molecular phenomena giving rise to the spectrum (such as rotation or vibration). Since we are concerned with interactions at the microscopic level, it is not surprising that the ideas of quantum mechanics are important in the development of this subject. In mathematical terms, most aspects of spectroscopy can be dealt with using techniques we have already met.

8.2 CALCULATION OF DIPOLE MOMENTS

For a heteronuclear diatomic molecule, with partial charge q on each atom and an interatomic separation of r (Figure 8.1), the dipole moment μ is given by

$$\mu = qr$$

Worked Example 8.1

Calculate the dipole moment of the methylidyne radical CH, given that the excess charge on each atom is $0.27e$ and the bond length 1.120 Å. Here e represents the electronic charge.

CHEMICAL BACKGROUND

For most second-row hydrides, the hydrogen atom will denote the positive end of the dipole. The exceptions are LiH and BeH, and in these cases, the dipole would be given a negative value. Methylidine was one of the first interstellar molecules to be identified and is involved in the creation of other organic molecules.

FIGURE 8.1 Charges of magnitude q separated by distance r forming a dipole.

DOI: 10.1201/9781003043218-8

Partial charges cannot be measured directly, but they can be determined by quantum mechanical calculations at various levels of sophistication. The resulting dipole moment is then an indication of their accuracy.

Dipole moments are usually given in units of debye (D). The debye is a non-SI unit which is equivalent to 3.336×10^{-30} C m.

Solution to Worked Example 8.1

Notice first of all that the excess charge q is given in terms of the electronic charge e which has the value 1.602×10^{-19} C. Therefore

$$q = 0.27 \times 1.602 \times 10^{-19}\,C = 4.325 \times 10^{-20}\,C$$

At this point, we will carry one more figure than is strictly justified in order to avoid intermediate rounding errors.

The interatomic distance is given in terms of the non-SI unit Å, where $1\ \text{Å} = 10^{-10}$ m. We now have

$$\mu = qr$$

$$= 4.325 \times 10^{-20}\,C \times 1.120 \times 10^{-10}\,m$$

$$= 4.844 \times 10^{-30}\,C\,m$$

To express this answer in D, we rearrange

$$1\,D = 3.335\,64 \times 10^{-30}\,C\,m$$

by dividing both sides by $3.335\,64 \times 10^{-30}$ so that

$$1\,C\,m = \frac{1}{3.335\,64 \times 10^{-30}}\,D$$

$$= 2.998 \times 10^{29}\,D$$

Consequently

$$\mu = 4.844 \times 10^{-30} \times 2.998 \times 10^{29}\,D$$

$$= 1.452\,D$$

Returning to the original data, we see that the lowest number of significant figures is 2. Since all the arithmetic operations have been multiplication and division, the final value of μ should be given to 2 significant figures. The value to quote is therefore 1.5 D.

8.3 DIPOLE AND QUADRUPOLE MOMENTS

The total electric potential $\phi(r)$ for the charge distribution a distance r and angle θ from the axis of a linear molecule is given by

$$\phi(r) = \frac{q}{4\pi\varepsilon_o r} + \frac{p\cos\theta}{4\pi\varepsilon_o r^2} + \frac{\Theta(3\cos^2\theta - 1)}{8\pi\varepsilon_o r^3} + \dots$$

where q is the net charge, p the dipole moment, and Θ the quadrupole moment. The constant ε_o is the permittivity of free space.

If the linear molecule consists of point charges then

$$p = \Sigma_i q_i z_i \text{ and } \Theta = \sum_i q_i z_i^2$$

where z_i represents distance along the molecular axis, measured from the centre of mass.

Worked Example 8.2

Calculate the dipole and quadrupole moments for HF which has a bond length 0.917 Å and partial charges 0.44e.

CHEMICAL BACKGROUND

The fact that HF is quite polar is reflected in the relatively high value of the partial charges. These will be equal and opposite because HF is a neutral molecule.

Solution to Worked Example 8.2

It is straightforward to convert the partial charges to SI units by multiplying them by the electronic charge e. This gives

$$q = 0.44 \times 1.602 \times 10^{-19} \, C$$

$$= 7.049 \times 10^{-20} \, C$$

The appropriate values of z are less straightforward to determine. Since the relative atomic masses of hydrogen and fluorine are 1.008 and 18.998 respectively we need to divide the interatomic distance 0.917 Å in this ratio as shown in Figure 8.2. To do this we add the masses to give 20.006. The centre of mass will be closer to the heavier fluorine atom, so

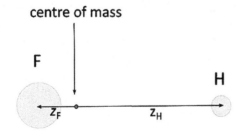

FIGURE 8.2   Centre of mass of the HF molecule.

$$z_F = \frac{1.008}{20.006} \times 0.917\,\text{Å}$$

$$= 0.046\,2\,\text{Å}$$

and

$$z_H = \frac{18.998}{20.006} \times 0.917\,\text{Å}$$

$$= 0.870\,8\,\text{Å}$$

As z_F and z_H are measured in opposite directions we actually set $z_H = -0.046\,2$ Å.

The dipole moment is given by

$$p = \Sigma_i q_i z_i$$

$$= q_H z_H + q_F z_F$$

$$= (7.049 \times 10^{-20}\,\text{C} \times 0.870\,8 \times 10^{-10}\,\text{m}) + (-7.049 \times 10^{-20}\,\text{C}) \times (-0.046\,2 \times 10^{-10}\,\text{m})$$

$$= 6.138 \times 10^{-30}\,\text{C m} + 0.326 \times 10^{-30}\,\text{C m}$$

$$= 6.464 \times 10^{-30}\,\text{C m}$$

This can be converted to units of D, as in the previous example, to give

$$6.464 \times 10^{-30}\,\text{C m} \times 2.998 \times 10^{29}\,\text{D}$$

$$= 1.94\,\text{D}$$

Notice that in calculating the above, we have applied the precedence rules as introduced in Section 1.3 on page 6.

The quadrupole moment is given by

$$\Theta = \sum_i q_i z_i^2$$

$$= q_H z^2{}_H + q_F z^2{}_F$$

$$= 7.049 \times 10^{-20}\,\text{C} \times (0.870\,8 \times 10^{-10}\,\text{m})^2 + (-7.049 \times 10^{-20}\,\text{C}) \times (-0.046\,2 \times 10^{-10}\,\text{m})^2$$

$$= 5.345 \times 10^{-40}\,\text{C m}^2 - 0.015\,05 \times 10^{-40}\,\text{C m}^2$$

$$= 5.330 \times 10^{-40}\,\text{C m}^2$$

again using the precedence rules.

8.4 ELECTROMAGNETIC RADIATION

Electromagnetic radiation is so called because it consists of an oscillating electric field and an oscillating magnetic field at right angles to one another. The electromagnetic spectrum ranges from gamma rays with a wavelength of about 10^{-12} m through to radio waves whose wavelength is approximately 100 m. In between, in ascending order of wavelength, are X-rays, ultraviolet radiation, visible light, infra-red radiation, microwaves and television waves (Figure 8.3). As well as being characterized by wavelength λ, electromagnetic radiation can be expressed in terms of its frequency ν.

8.4.1 DIRECT AND INVERSE PROPORTION

These types of proportions are discussed in Section 3.7.4 on page 95.

Worked Example 8.3

The relationship between the wavelength λ and frequency ν of electromagnetic radiation is

$$c = \lambda \nu$$

where c is its velocity, which can be taken as a constant. How can the relationship between λ and ν be described?

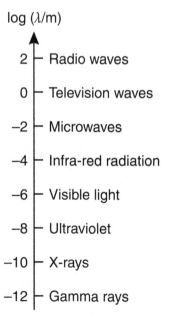

FIGURE 8.3 The electromagnetic spectrum.

CHEMICAL BACKGROUND

The energy E of a photon is given by the equation

$$E = h\nu$$

where h is Planck's constant. Therefore, as the frequency of the radiation increases so does its energy. Since the wavelength is inversely proportional to the frequency, it also follows that higher wavelength radiation will have a lower energy. An example is that of ultraviolet light which has a lower energy and frequency than that of visible light. Consequently, it is more effective in inducing photochemical reactions.

The velocity of light is exactly 299 792 458 m s^{-1}. As well as being measured experimentally, this value can be predicted from Maxwell's equations which describe the behaviour of electric and magnetic fields.

Solution to Worked Example 8.3

Since the product of the two variables is a constant (the velocity of light), we should be able to recognize instantly that they are inversely proportional. Alternatively, we can rearrange the equation given, by dividing both sides by λ, so that λ appears on one side and ν on the other.

$$\frac{c}{\lambda} = \nu$$

It should now be apparent that as we increase λ then ν must decrease.

Worked Example 8.4

In addition to wavelength and frequency, a quantity often used in spectroscopy is wavenumber. This is normally given the symbol $\tilde{\nu}$, defined as the reciprocal of the wavelength, i.e.

$$\tilde{\nu} = \frac{1}{\lambda}$$

and has the units cm^{-1}. What is the wavenumber of the sodium D line having frequency 5.086×10^8 MHz? Take the velocity of electromagnetic radiation to be 2.998×10^8 m s^{-1}.

CHEMICAL BACKGROUND

When excited by an electric discharge, sodium vapour produces an emission spectrum which includes a yellow line at a wavelength of 589 nm. On closer analysis, this is seen to consist of two very closely spaced lines (a doublet) of wavelengths 589.76 and 589.16 nm respectively.

This type of discharge is typically seen in certain forms of street lighting.

Solution to Worked Example 8.4

We need to calculate the value of the quantity $\frac{1}{\lambda}$. If we start with the defining equation

$$c = \lambda v$$

we can divide both sides by v to give

$$\frac{c}{v} = \lambda$$

As we wish to calculate $\frac{1}{\lambda}$, we now need to take the reciprocal of each side of this equation. The reciprocal of λ is $\frac{1}{\lambda}$ while the reciprocal of any fraction is obtained by 'turning it upside down'. This now gives us

$$\frac{1}{\lambda} = \frac{v}{c}$$

$$= \frac{5.086 \times 10^8 \, \text{MHz}}{2.998 \times 10^8 \, \text{m s}^{-1}}$$

The first thing to notice is that both the top and the bottom of this fraction contain a 10^8 term, which can be cancelled. We then need to rewrite the unit MHz in a form which is of more use to us. The basic frequency unit Hz is equal to s^{-1}, while the prefix M means 10^6 (Appendix A). This now gives

$$\frac{1}{\lambda} = \frac{5.086 \times 10^6 \, \text{s}^{-1}}{2.998 \, \text{m s}^{-1}}$$

The s^{-1} unit now appears in both the top and bottom of the right side of this equation and can be cancelled. Using a calculator we now obtain

$$\frac{1}{\lambda} = 1.696 \times 10^6 \, \text{m}^{-1}$$

While this is an acceptable unit for wavenumber, it is more usual to use units of cm^{-1}. We can perform the necessary calculation if we realize that

$$100 \, \text{cm} = 1 \, \text{m}$$

or more usefully here

$$10^2 \, \text{cm} = 1 \, \text{m}$$

We replace m in our expression with 10^2 cm. This gives

$$\frac{1}{\lambda} = 1.696 \times 10^6 \, m^{-1}$$

$$= 1.696 \times 10^6 \times (10^2 \, cm)^{-1}$$

The power of −1 is applied to every term in the bracket, so we need to use the fact that $(10^2)^{-1} = 10^{-2}$ to give

$$\frac{1}{\lambda} = 1.696 \times 10^6 \times 10^{-2} \, cm^{-1}$$

$$= 1.696 \times 10^4 \, cm^{-1}$$

8.5 THE BEER-LAMBERT LAW

When radiation is incident upon some medium, it will be absorbed to a certain extent, as shown in Figure 8.4. This will result in its intensity being reduced, and such behaviour is described by the Beer-Lambert Law which can be expressed by the equation

$$A = \varepsilon c l$$

where ε is a constant known as the absorption coefficient, c is the concentration of the absorbing medium and l is the distance through which the light travels within the medium. The quantity A is known as the absorbance and is defined as

$$A = \log \frac{I_o}{I}$$

where I_o is the intensity of the incident radiation and I is the intensity once it has passed through the medium. The distance l is commonly known as the path length.

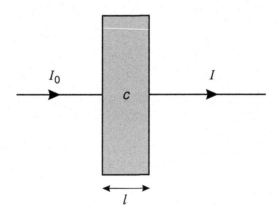

FIGURE 8.4   The absorption of radiation.

8.5.1 RULES OF LOGARITHMS

These were discussed in Section 5.5.2 on page 161.

Worked Example 8.5

The fraction of incident radiation transmitted is the transmittance T which is defined by the equation

$$T = \frac{I}{I_o}$$

Obtain an equation which relates the percentage transmittance T_{100} to the absorbance A.

CHEMICAL BACKGROUND

Many spectrophotometers are capable of displaying both absorbance and transmittance, and the relationship between these quantities is not immediately obvious. Generally, absorbance is the preferred quantity because of its proportionality to both concentration and path length.

Spectrophotometers are generally designed to work in one region of the spectrum only. Specific wavelengths are produced by using a rotating prism or a diffraction grating.

Solution to Worked Example 8.5

The percentage transmittance T_{100} is related to the transmittance T (a fractional value) by the simple expression

$$T_{100} = 100T$$

or

$$T_{100} = 100\frac{I}{I_o}$$

This can be rearranged by dividing by T_{100} and multiplying by $\frac{I_o}{I}$ to give

$$\frac{I_o}{I} = \frac{100}{T_{100}}$$

which can be substituted into the expression for A

$$A = \log\frac{I_o}{I}$$

$$= \log\frac{100}{T_{100}}$$

Since we know that

$$\log\left(\frac{X}{Y}\right) = \log X - \log Y$$

we can rearrange the right-hand side of this equation to give

$$A = \log 100 - \log T_{100}$$

and since $\log 100 = \log 10^2 = 2$ (as we saw in Section 3.6.1 on page 80) we obtain

$$A = 2 - \log T_{100}$$

which finally can be rearranged by adding T_{100} and subtracting A from both sides to give

$$\log T_{100} = 2 - A$$

8.6 ROTATIONAL SPECTROSCOPY

Rotating molecules give rise to spectra in the microwave region. It is possible to define three mutually perpendicular principal axes of rotation which pass through the centre of gravity of a molecule, and there are consequently three corresponding moments of inertia which depend on the molecular shape. The treatment of rotational spectra is consequently different for linear molecules (such as HCl), symmetric tops (such as CH_3F), spherical tops (such as CH_4) and asymmetric tops (such as H_2O). These different types of molecules are shown in Figure 8.5.

8.6.1 SEQUENCES

This topic was introduced in Section 7.5.2 on page 252.

Worked Example 8.6

The rotational energy levels $F(J)$ of a rigid diatomic molecule are given by the expression

$$F(J) = BJ(J + 1)$$

where J is the rotational quantum number (taking values 0, 1, 2,...) and B is the rotational constant. What are the three lowest rotational energy levels, in terms of B?

CHEMICAL BACKGROUND

This problem shows that the rotational energy levels in a diatomic molecule are not evenly spaced. However, if we consider the

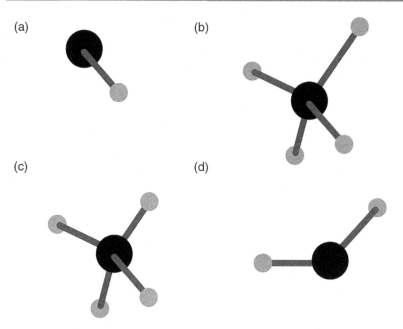

(a)

(b)

(c)

(d)

FIGURE 8.5 Molecules classified according to their rotational properties: (a) HCl (linear); (b) CH$_3$F (symmetric top); (c) CH$_4$ (spherical top); and (d) H$_2$O (asymmetric top).

differences between levels then we do see an even spacing, and this is apparent in rotational spectra.

Each energy level has a degeneracy of $(2J + 1)$; in other words, there are $(2J + 1)$ levels having each energy value. The population of each level of energy $F(J)$ is proportional to $(2J+1)\exp\left(-\dfrac{F(J)}{kT}\right)$.

Solution to Worked Example 8.6

To obtain the three lowest energy levels, we need to substitute the three lowest values of J into the given expression. We then have

$$F(0) = (B \times 0)(0 + 1) = 0$$

$$F(1) = (B \times 1)(1 + 1) = 2B$$

$$F(2) = (B \times 2)(2 + 1) = 6B$$

as shown in Figure 8.6.

Worked Example 8.7

The rotational energy levels $F(J)$ *of* a rigid diatomic molecule are given by the expression

$$F(J) = BJ(J + 1)$$

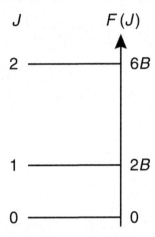

FIGURE 8.6   Spacing of rotational energy levels.

where J is the rotational quantum number (0,1,2,...) and B is the rotational constant. Obtain an expression for the energy of a rotational transition, given that such transitions are subject to the selection rule $\Delta J = \pm 1$.

CHEMICAL BACKGROUND

The rotational constant B can be calculated from the equation

$$B = \frac{h^2}{8\pi^2 I}$$

where h is Planck's constant and I is the moment of inertia. The selection rule is obtained from the application of the Schrödinger equation.

Solution to Worked Example 8.7

To calculate the energy of a transition, we need to obtain expressions for the energies of two adjacent levels and then determine the difference between these. We choose these to have rotational quantum numbers J and $J + 1$ respectively. We then have

$$F(J) = BJ(J + 1)$$

as given above. This can be expanded to give

$$F(J) = B(J^2 + J)$$

To obtain $F(J + 1)$, we need to replace each occurrence of J in the above equation by $J + 1$. This gives

$$F(J + 1) = B(J + 1)((J + 1) + 1)$$

$$= B(J + 1)(J + 2)$$

We multiply the two brackets together, as shown in Section 3.4.1 on page 64, by considering the four combinations of terms to obtain

$$F(J+1) = B(J^2 + 2J + J + 2)$$

$$= B(J^2 + 3J + 2)$$

The energy difference $\Delta F(J)$ is given by

$$\Delta F(J) = F(J+1) - F(J)$$

$$= B(J^2 + 3J + 2) - B(J^2 + J)$$

$$= B(J^2 + 3J + 2 - J^2 - J)$$

$$= B(2J + 2)$$

We can remove the common factor of 2 from within the bracket to leave

$$\Delta F(J) = 2B(J + 1)$$

8.7 VIBRATIONAL SPECTROSCOPY

The starting point for a study of vibrational spectroscopy is usually the model of a simple harmonic oscillator, a system which is often treated in some detail in courses on quantum mechanics. We met this in Problem 5 of Chapter 7 on page 284. Any energy possessed by this system over and above its equilibrium value is due to extension or compression of the bond. The vibrational frequency, ω_e, of this system is given by the formula

$$\omega_e = \frac{1}{2\pi}\sqrt{\frac{k}{\mu}}$$

where k is the force constant and μ is a quantity known as the reduced mass of the system, which is related to the masses of the atoms forming the bonds.

Worked example 8.8

The vibrational energy $G(v)$ of the harmonic oscillator is given by the equation

$$G(v) = \left(v + \frac{1}{2}\right)\omega_e$$

where v is the vibrational quantum number (having values 0, 1, 2,...). Calculate the energy of the vibrational level having $v = 2$ for the HCl molecule. Its force constant is 516 N m^{-1} and its reduced mass is given by the equation

$$\mu = \frac{m_{Cl} m_H}{m_{Cl} + m_H}$$

where m represents the mass of the specified atom.

One way of remembering the correct formula for calculating the reduced mass is to consider the units. In the formula here, we have units of mass squared divided by units of mass; the reciprocal would clearly give incorrect units.

CHEMICAL BACKGROUND

In practice, there will be very few molecules in this vibrational level, so it will not contribute significantly to the vibrational spectrum.

Solution to Worked Example 8.8

For 1 mol of HCl, we would have

$$\mu = \frac{35.453\,\text{g mol}^{-1} \times 1.007\,9\,\text{g mol}^{-1}}{35.453\,\text{g mol}^{-1} + 1.007\,9\,\text{g mol}^{-1}}$$

$$= \frac{35.733\,\text{g mol}^{-1}}{36.461}$$

$$= 0.980\,03\,\text{g mol}^{-1}$$

and for one molecule we need to divide by Avogadro's constant:

$$\mu = \frac{0.980\,03\,\text{g mol}^{-1}}{6.022 \times 10^{23}\,\text{mol}^{-1}}$$

$$= \frac{0.162\,74\,\text{g mol}^{-1}}{10^{23}\,\text{mol}^{-1}}$$

$$= 1.627\,4 \times 10^{-24}\,\text{g}$$

$$= 1.627\,4 \times 10^{-27}\,\text{kg}$$

so from the formula above the fundamental frequency can be calculated as

$$\omega_e = \frac{1}{2\pi}\sqrt{\frac{k}{\mu}}$$

$$= \frac{1}{2\pi}\sqrt{\frac{516\,\text{N m}^{-1}}{1.627\,4 \times 10^{-27}\,\text{kg}}}$$

Since 1 N = 1 kg m s^{-2} this becomes

$$\omega_e = \frac{1}{2\pi}\sqrt{\frac{516\,\text{kg m s}^{-2}\,\text{m}^{-1}}{1.627\,4 \times 10^{-27}\,\text{kg}}}$$

Cancelling units (kg top and bottom, and m and m^{-1}) then gives

$$\omega_e = \frac{1}{2\pi}\sqrt{\frac{516\,\text{s}^{-2}}{1.627\,4\times10^{-27}}}$$

$$= \frac{\sqrt{3.171\times10^{29}\,\text{s}^{-2}}}{2\pi}$$

$$= \frac{5.631\times10^{14}\,\text{s}^{-1}}{2\pi}$$

$$= 8.962\times10^{13}\,\text{s}^{-1}$$

We now need to substitute the value of $v = 2$ into the expression for the vibrational energy. This gives us

$$G(2) = \left(2+\frac{1}{2}\right)\omega_e$$

$$= 2.5\times8.962\times10^{13}\,\text{s}^{-1}$$

$$= 2.24\times10^{14}\,\text{s}^{-1}$$

$$= 2.24\times10^{14}\,\text{Hz}$$

since $1\,\text{Hz} = 1\,\text{s}^{-1}$ and we need to restrict our answer to 3 significant figures as given in the original data.

This answer is easily transformed into other quantities by the use of appropriate conversion factors. Multiplication by Planck's constant h will give energy (in J) and division by the velocity of light c will give wavenumber (in m^{-1}, easily converted to cm^{-1}).

Worked Example 8.9

The vibrational energy $G(v)$ of the harmonic oscillator is given by the equation

$$G(v) = \left(v+\frac{1}{2}\right)\omega_e$$

where v is the vibrational quantum number (0, 1, 2,...). Given that the selection rule for a vibrational transition is $\Delta v = \pm 1$, obtain an expression for the energy associated with such a transition.

CHEMICAL BACKGROUND

The energy associated with this transition turns out to be a constant value, equal to the vibrational frequency of the system. This is because the vibrational energy levels are equally spaced and the vibrating molecule will only absorb energy of the same frequency as its own natural vibrations.

The harmonic oscillator approximation is valid for vibrations which displace the bond length by up to about 10% of its equilibrium value. Above that, it is better to use the model of an anharmonic oscillator, as used in Problem 8 at the end of this chapter.

Solution to Worked Example 8.9

As in the rotational case in Worked example 8.7, we need to set up expressions for the energy of two adjacent levels. We have from the question

$$G(v) = \left(v + \frac{1}{2} \right) \omega_e$$

Replacing every occurrence of v by $v + 1$ gives

$$G(v+1) = \left((v+1) + \frac{1}{2} \right) \omega_e$$

$$= \left(v + 1 + \frac{1}{2} \right) \omega_e$$

$$= \left(v + \frac{3}{2} \right) \omega_e$$

The energy $\Delta G(v)$ associated with this vibrational transition will now be given by

$$\Delta G(v) = G(v+1) - G(v)$$

$$= \left(v + \frac{3}{2} \right) \omega_e - \left(v + \frac{1}{2} \right) \omega_e$$

$$= \left(v + \frac{3}{2} - v - \frac{1}{2} \right) \omega_e$$

$$= \omega_e$$

The spacing of vibrational energy levels is shown in Figure 8.7.

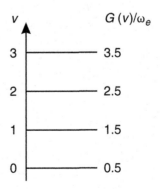

FIGURE 8.7   Spacing of vibrational energy levels.

8.8 ROTATION-VIBRATION SPECTROSCOPY

It is reasonable to suppose that a molecule will be undergoing both rotational and vibrational motion at a given time and it is possible to investigate the coupling of these motions by means of spectroscopic techniques. We assume that it is possible to separate the vibrational and rotational motion, and it is therefore possible to add the contributions from each to obtain the vibration-rotation energy. Figure 8.8 shows the rotational levels superimposed on the vibrational levels for a molecule.

Worked Example 8.10

The energy $E(v, J)$ due to combined vibration and rotation is given by the equation

$$E(v,J) = BJ(J+1) + \left(v + \frac{1}{2}\right)\omega_e$$

where v and J are the vibrational and rotational quantum numbers respectively. Given the selection rules $\Delta v = \pm 1$ and $\Delta J = \pm 1$, obtain an expression for the energy due to simultaneous vibrational and rotational transitions.

CHEMICAL BACKGROUND

The selection rules given show that it is only possible for a vibrational change to occur if a rotational change takes place simultaneously. It is possible to have $\Delta v = 0$, which represents the pure rotational change we met in Worked Example 7.4, but it is not possible to have $\Delta J = 0$ representing a vibrational change alone.

It is important to realize that the separation of the rotational levels is much less than that of the vibrational levels. Typical values are 0.05 kJ mol^{-1} for rotational levels and 10 kJ mol^{-1} for vibrational levels. These compare with a separation of the order of 500 kJ mol^{-1} for electronic levels.

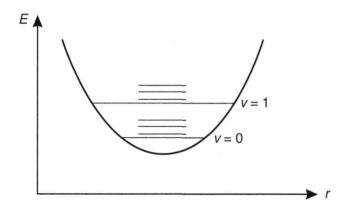

FIGURE 8.8   Vibrational and rotational energy levels on the same scale.

Solution to Worked Example 8.10

We proceed as in the previous problems by considering the difference between two energy levels, remembering this time that we need to deal with two simultaneous changes. The notation usually employed is that representing a transition from a lower vibrational level v'' and rotational level J'' to an upper vibrational level v' and corresponding rotational level J. We then have

$$\Delta E(v, J) = E(v', J') - E(v'', J'')$$

$$= BJ'(J'+1) + \left(v' + \frac{1}{2}\right)\omega_e - BJ''(J''+1) - \left(v'' + \frac{1}{2}\right)\omega_e$$

$$= BJ'(J'+1) - BJ''(J''+1) + (v' - v'')\omega_e$$

Since we have the selection rule $\Delta v = \pm 1$ it follows that

$$v' - v'' = 1$$

since v' is the higher level and so

$$\Delta E(v, J) = BJ'(J'+1) - BJ''(J''+1) + \omega_e$$

When we move to the higher vibrational level, it is possible for the change in rotational quantum number to be either positive or negative. This is a consequence of the rotational energy levels being more closely spaced than the vibrational energy levels If

$$J' = J'' + 1$$

and therefore

$$J'' = J' - 1$$

we have

$$\Delta E(J') = BJ'(J'+1) - B(J'-1)J' + \omega_e$$

$$= BJ'^2 + BJ' - BJ'^2 + BJ' + \omega_e$$

$$= \omega_e + 2\,BJ'$$

Alternatively, if

$$J' = J'' - 1$$

then

$$\Delta E(J'') = B(J''-1)J'' - BJ''(J''+1) + \omega_e$$

$$= BJ''^2 - BJ'' - BJ''^2 - BJ'' + \omega_e$$

$$= \omega_e - 2BJ''$$

Note that either of these expressions could be obtained using J' or J'.

8.9 NUCLEAR MAGNETIC RESONANCE SPECTROSCOPY

This form of spectroscopy is one of the most powerful analytical tools used by organic chemists. While spectra for complicated molecules can be quite difficult to unravel, the underlying mathematics is quite simple to understand. Such spectra are basically due to the splitting of the proton spin energy levels by a magnetic field. Peaks are classified by their chemical shifts (measured in parts per million, ppm) relative to the standard tetramethylsilane, $SiMe_4$.

Chemical shifts can be measured on the τ scale, on which tetramethylsilane has a value of 10. The chemical shifts of most groups of interest then fall between 0 and 10, with a phenyl hydrogen having a value of 2.7, for example.

8.9.1 PASCAL'S TRIANGLE

If we wish to expand the expression $(x + 2)^2$ we could rewrite it as

$$(x + 2)^2 = (x + 2)(x + 2)$$

and multiply out the brackets to give

$$(x + 2)^2 = x(x + 2) + 2(x + 2)$$

$$= x^2 + 2x + 2x + 4$$

$$= x^2 + 4x + 4$$

Similarly, $(x + 2)^3$ could be calculated as

$$(x + 2)^3 = (x + 2)(x + 2)^2$$

$$= (x + 2)(x^2 + 4x + 4)$$

$$= x(x^2 + 4x + 4) + 2(x^2 + 4x + 4)$$

$$= x^3 + 4x^2 + 4x + 2x^2 + 8x + 8$$

$$= x^3 + 6x^2 + 12x + 8$$

Similarly, we could obtain expressions for $(x + 2)^4$ and higher powers, but this is obviously going to become more and more difficult as the power increases.

Fortunately, such step-by-step expansion is not necessary if we use a device known as **Pascal's Triangle.** This is shown in Figure 8.9. You will probably be relieved to know that this does not need to be memorized, as it can be set up using a few simple rules which are easily remembered.

We start on the top line with the number 1.

1

This will actually be of more use to us if we write it with zeros on either side as

$$0 \quad 1 \quad 0$$

To obtain the second line, we now add each pair of numbers in the line above and write the result in the space underneath them. This gives

$$
\begin{array}{ccc}
0 & 1 & 0 \\
 & + \quad + & \\
 & 1 \quad 1 &
\end{array}
$$

We now add the zeros on either side of this second line:

$$
\begin{array}{cccc}
 & 0 & 1 & 0 \\
0 & 1 & & 1 \quad 0
\end{array}
$$

To obtain the third line, we repeat the process of adding each pair of numbers in the second line and writing the result in the space underneath. We then have

$$
\begin{array}{ccccc}
 & & 0 & 1 & 0 \\
 & 0 & 1 & & 1 \quad 0 \\
 & + & & + & + \\
0 & 1 & 2 & 1 & 0
\end{array}
$$

The rest of Pascal's Triangle can be built up in the same way.

How does this help us to expand brackets to a specified power? From the examples given, we can see that when a bracket such as $(x + 2)$ is raised to some power, we obtain a polynomial expression. These were discussed in Section 3.4.2 on page 65, where we saw that each term in x is multiplied by a constant called the coefficient. For example, in a term such as $3x^4$, we say that the coefficient of x^4 is 3. It is these coefficients that can be determined by using Pascal's Triangle.

From Figure 8.9, we see that the first line of Pascal's Triangle gives us the coefficient of a zero-order expansion (power 0), the next line a

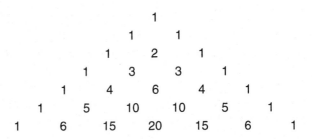

FIGURE 8.9   Pascal's triangle.

first-order expansion (power 1) and so on. Therefore to evaluate $(x + 2)^2$ we need the coefficients from the third line, which are 1, 2, 1. As we saw earlier, we would expect this expansion to give us terms in x^2, x and a constant term, so we have

$$\text{coefficient of } x^2 = 1$$

$$\text{coefficient of } x = 2$$

$$\text{coefficient of constant} = 1$$

This expansion can be expressed as

$$(x + 2)^2 = 1 \times x^2 \times 2^0 + 2 \times x^1 \times 2^1 + 1 \times x^0 \times 2^2$$

where each term comprises the appropriate coefficient from Pascal's Triangle, a term in x and a term in 2. Notice that as the power of x decreases that of 2 increases, but the two powers added together are always equal to 2. If we remember that any number raised to a zero power is 1 (Section 1.6 on page 16), and any number raised to power 1 is itself, this expression simplifies to give

$$(x + 2)^2 = 1 \times x^2 \times 1 + 2 \times x \times 2 + 1 \times 1 \times 4$$

$$= x^2 + 4x + 4$$

which is, of course, the result we obtained before.

Let us now consider an example which would be much more difficult to evaluate without Pascal's Triangle, $(2x + 3)^4$. From Figure 8.9 we see that the appropriate coefficients for an expansion of degree 4 are

$$1\ 4\ 6\ 4\ 1$$

This time we need to consider terms in $2x$ and 3, so we have

$$(2x + 3)^4 = 1 \times (2x)^4 \times 3^0 + 4 \times (2x)^3 \times 3^1$$

$$+ 6 \times (2x)^2 \times 3^2 + 4 \times (2x)^1 \times 3^3 + 1 \times (2x)^0 \times 3^4$$

Notice that when we raise a bracket to the fourth power we have five terms. The power of $(2x)$ decreases while that of 3 increases and the sum of these powers is always 4. It is important to realize that, for example,

$$(2x)^4 = 2^4 x^4 = 16x^4$$

so the power is applied to *both* quantities within the bracket. The expansion can then be simplified to

$$(2x + 3)^4 = (1 \times 16x^4 \times 1) + (4 \times 8x^3 \times 3) + (6 \times 4x^2 \times 9)$$

$$+ (4 \times 2x \times 27) + (1 \times 1 \times 81)$$

$$= 16x^4 + 96x^3 + 216x^2 + 72x + 81$$

Worked Example 8.11

The peaks due to each group of protons in a molecule are split into $n + 1$ by an adjacent group of n protons, the intensity of these peaks being given by the coefficients of Pascal's Triangle. What is the distribution of peaks within the CH_2 and CH_3 signals of ethanol (Figure 8.10), ignoring any interaction with the OH group?

CHEMICAL BACKGROUND

The reason for ignoring any interaction with the OH group is that its proton undergoes rapid exchange with other molecules, due to the presence of small amounts of acid, and so it is unable to interact with protons in the other groups of the molecule.

Solution to Worked Example 8.11

The signal due to the CH_3 group will be split into three $(2 + 1)$ peaks by the CH_2 group, and from Pascal's Triangle these will be in the ratio 1:2: 1. The signal due to the CH_2 group will be split into four $(3 + 1)$ peaks by the CH_3 group, in the ratio 1: 3: 3: 1. These peaks are shown in Figure 8.11.

Nuclear magnetic resonance spectra are most commonly measured for the ^1H nucleus. Other nuclei which can be used include ^{19}F, ^{31}P and, since the advent of Fourier transform methods, ^{13}C.

OH

FIGURE 8.10   Structure of ethanol.

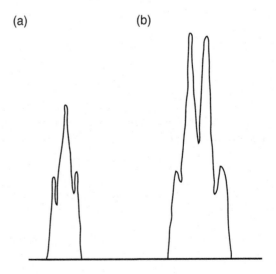

FIGURE 8.11   Nuclear magnetic resonance spectrum of ethanol: (a) triplet due to CH_3 group; and (b) quartet due to CH_2 group.

8.10 FOURIER TRANSFORM SPECTROSCOPY

Fourier transforms are used in both infrared and nuclear magnetic resonance spectroscopy. Essentially they allow conversion between wave-
forms and spectra, allowing the analysis of a complicated spectrum to
give its components in terms of individual frequencies. The spectrometers
used contain a built-in computer which performs the necessary integrations. A further use for Fourier transforms is in X-ray crystallography.

8.10.1 INTRODUCTION TO FOURIER TRANSFORMS

If two functions $f(x)$ and $g(y)$ are related by the equations

$$f(x) = \int_{-\infty}^{\infty} g(y)e^{ixy}\,dy$$

and

$$g(y) = \frac{1}{2\pi}\int_{-\infty}^{\infty} f(x)e^{-ixy}\,dx$$

then we say that $g(y)$ is the Fourier transform of $f(x)$ and that $f(x)$ is the
inverse Fourier transform of $g(y)$. Note that i is the imaginary number
such that $i^2 = -1$ as introduced in Section 6.6.1 on page 224. You will
see various equations used to define these Fourier transform pairs. These
are likely to differ in the placing of a factor of 2π and the position of
positive and negative exponentials. It is important to be aware of whatever definition is used in a particular case. The one above allows us to
explore the mathematical techniques that are required regardless of the
definition in a particular case.

When determining Fourier transforms we frequently have discontinuous functions; in these cases, the form of the function varies according to
the value of the variable. For example, a function $f(x)$ could be defined as

$$f(x) = \begin{cases} 0 \text{ for } x < 0 \\ x \text{ for } 0 < x < 1 \\ 0 \text{ for } x > 1 \end{cases}$$

as shown in Figure 8.12.

The Fourier transform of $f(x)$ is then defined by

$$g(y) = \frac{1}{2\pi}\int_{0}^{1} xe^{-ixy}\,dx$$

these particular limits being specified because the integrals for $x < 0$ and
$x > 1$ will be zero. Since the remaining integral is with respect to the

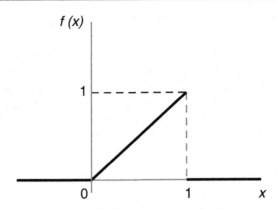

FIGURE 8.12   The function $f(x)$ defined in Section 8.10.1.

variable x, we treat y as constant and therefore need to use the standard integral

$$\int xe^{ax}\,dx = \frac{e^{ax}}{a^2}(ax-1)+C$$

where $a = -iy$. Consequently

$$g(y) = \frac{1}{2\pi}\left[\frac{e^{-ixy}}{(-iy)^2}(-ixy-1)\right]_0^1$$

This can be simplified because

$$(-iy)^2 = i^2y^2$$

$$= -y^2$$

since $i^2 = -1$ so that

$$g(y) = \frac{1}{2\pi}\left[\frac{e^{-ixy}}{-y^2}(-ixy-1)\right]_0^1$$

If we multiply the top and bottom of this fraction by -1, we then have

$$g(y) = \frac{1}{2\pi}\left[\frac{e^{-ixy}}{y^2}(1+ixy)\right]_0^1$$

Substituting for the limits in x gives us

$$g(y) = \frac{1}{2\pi}\left\{\left[\frac{e^{-iy}}{y^2}(1+iy)\right]-\left[\frac{e^0}{y^2}\right]\right\}$$

$$= \frac{1}{2\pi}\left[\frac{e^{-iy}}{y^2}(1+iy)-\frac{1}{y^2}\right]$$

since $e^0 = 1$ and so

$$g(y) = \frac{1}{2\pi y^2}\left[e^{-iy} + iye^{-iy} - 1\right]$$

This can be further simplified to extract the real and imaginary parts of $g(y)$.

Worked Example 8.12

In Fourier transform spectroscopy, radiation can be applied to a sample in terms of a pulse. This can be described in terms of a function of time $f(t)$, for example

$$f(t) = \begin{cases} 0 \text{ for } t < 1 \\ 2 \text{ for } 1 < t < 2 \\ 0 \text{ for } t > 2 \end{cases}$$

as shown in Figure 8.13. Determine the Fourier transform $g(\omega)$ in terms of frequency ω.

CHEMICAL BACKGROUND

This is a very simplistic representation of the application of radiation, but it does allow us to illustrate the use of the Fourier transform to move from a time domain to a frequency domain. A more realistic shape for the pulse of radiation would be a Gaussian distribution; in that case, the resulting Fourier transform is also a Gaussian-shaped normal distribution.

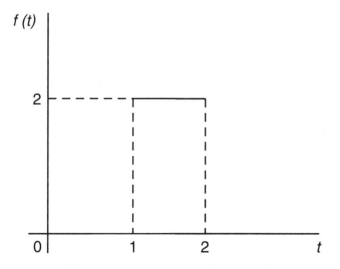

FIGURE 8.13   The function $f(x)$ defined in Worked Example 8.12.

Solution to Worked Example 8.12

The equations for the Fourier transform in terms of the variables t and ω become

$$f(t) = \int_{-\infty}^{\infty} g(\omega) e^{it\omega} d\omega$$

and

$$g(\omega) = \frac{1}{2\pi} \int_{-\infty}^{\infty} f(t) e^{-it\omega} dt$$

Consequently

$$g(\omega) = \frac{1}{2\pi} \int_{1}^{2} 2e^{-it\omega} dt$$

since $f(t)$ is only non-zero between these limits. This is a straightforward integration of the form

$$\int e^{ax} dx = \frac{e^{ax}}{a} + C$$

with $a = -i\omega$ and $x = t$, ω being constant as we are integrating with respect to t. Therefore,

$$g(\omega) = \frac{1}{2\pi} \left[2 \frac{e^{-it\omega}}{-it\omega} \right]_{1}^{2}$$

$$= \frac{1}{2\pi} \left[\frac{2e^{-2i\omega}}{-2i\omega} - \frac{2e^{-i\omega}}{-i\omega} \right]$$

$$= \frac{1}{2\pi i\omega} \left[2e^{-i\omega} - 2e^{-2i\omega} \right]$$

$$= \frac{1}{\pi i\omega} \left[e^{-i\omega} - e^{-2i\omega} \right]$$

EXERCISES

1. Given the values

i	1	2	3	4	5
a_i	2.3	1.8	0.6	4.2	5.1
x_i	−0.16	0.19	−0.24	0.54	0.88

Determine the values of $\sum_i a_i x_i$ and $\sum_i a_i x_i^2$.

2. If $s = \dfrac{1}{t}$ and $t = \dfrac{u}{v}$, calculate s if $u = 41.6$ and $v = 15.8$.

3. If z is defined by

$$z = \frac{xy}{x+y}$$

calculate z if $x = 14.9$ and $y = -3.2$.

4. Identify the type of proportion between the appropriate functions of x and y in the relationships:
   a. $x \ln y = K$
   b. $\dfrac{x^2}{y} = K$
   c. $xe^{-y} = K$
   d. $xe^{-y} = K$
   where K is a constant.

5. Identify the proportionality constant when y is written as a function of x for the relationships:
   a. $ax = by$
   b. $\left(\dfrac{x}{a}\right)^2 = \left(\dfrac{b}{y}\right)$
   c. $\dfrac{y}{\ln x} = \dfrac{a}{b}$
   d. $ye^x = \dfrac{a}{b}$

6. Write each of these expressions as a single logarithm.
   a. $\ln x^3 - \ln x^2$
   b. $\ln xy + \ln x^2$
   c. $\ln\left(\dfrac{x}{y}\right) - \ln(xy)$
   d. $\ln x^2y - \ln xy^2$

7. Write each of these expressions as a sum or difference of logarithms.
   a. $\ln\sqrt{\dfrac{x+1}{x-1}}$
   b. $\ln 5x^4$
   c. $\ln\left(\dfrac{3x^2y}{z}\right)$
   d. $\ln(xy\sqrt{z})$

8. Simplify the expressions:
   a. $\log(4x+2) - \log(2x+1)$
   b. $\log(x^2 - 2x) - \log(x-2)$
   c. $\log 5x^3 - \log x$
   d. $\log(x+1) + \log(x-1)$

9. The terms in a series are defined by the expression

$$f(n) = 3n^3 + 4n^2 - 2n + 1$$

What is the value of $f(2) - f(1)$?

10. The terms in a series are defined by the expression

$$g(N) = N^2 + 2N$$

Obtain an expression for $g(N+1) - g(N)$.

11. Verify that the sequence

$$1, \frac{3}{2}, \frac{7}{4}, \frac{15}{8}, \frac{31}{16}$$

can be represented as

$$f(n) = 2 - \frac{1}{2^{n-1}}$$

What is the limit of this sequence as n tends to infinity?

12. The sine function can be represented by a series in x by means of the equation:

$$\sin x = a + bx + cx^2$$

   a. Set $x = 0$ to determine the value of a.
   b. Differentiate this equation to give an expression for $\cos x$.
   c. Set $x = 0$ in the expression for $\cos x$ you obtained in (b) to determine the value of b.
   d. Repeat the process in steps (b) and (c) to determine the value of c.
   e. Hence predict the series expression for $\sin x$.

13. Use Pascal's Triangle to obtain expressions for
   a. $(x+3)^4$
   b. $(2x-3)^4$
   c. $(2x-4)^5$
   d. $(3x+2)^6$

14. Use Pascal's triangle to obtain expressions for:

   a. $(x+y)^5$

   b. $\left(\dfrac{3}{x}-y\right)^4$

   c. $(2x+5y)^4$

   d. $\left(\dfrac{2}{x}+\dfrac{3}{y}\right)^5$

15. In the expansion of $(x-3y)^4$, what is the coefficient of y^3?

16. Use Pascal's Triangle to obtain an expression for $(1-x)^6$. By setting x to 0.002, obtain the value of 0.998 6 to 6 significant figures *without* using a calculator.

17. Obtain a general expression for the sum of the numbers in each row n of Pascal's triangle.

18. The Fibonacci sequence

$$1, 1, 2, 3, 5, 8...$$

   is formed by starting with 1 and adding the two previous numbers. Verify that the Fibonacci sequence is given by the sums of the diagonals in Pascal's triangle, if the triangle is written with the first number in each line at the left margin.

19. Determine the Fourier transform $g(y)$ of

$$f(x)=\begin{cases}0 & \text{for } x<-\pi \\ \sin 3x & \text{for } -\pi<x<\pi \\ 0 & \text{for } x>\pi\end{cases}$$

   using the standard integral

$$\int e^{ax}\sin bx\,dx = \frac{e^{ax}}{a^2+b^2}(a\sin bx - b\cos bx)+C$$

20. Determine the Fourier transform of

$$f(x)=\begin{cases}0 & \text{for } x<0 \\ x^2 & \text{for } 0<x<1 \\ 0 & \text{for } x>1\end{cases}$$

using the standard integral

$$\int x^n e^{ax} dx = \frac{1}{a} x^n e^{ax} - \frac{n}{a} \int x^{n-1} e^{ax} dx$$

PROBLEMS

1. Calculate the dipole and quadrupole moments for HCl which has a bond length 1.31 Å and calculated partial charges of $0.673e$.

2. The bond length of KBr is 2.82 Å and its dipole moment is 10.5 D. What are the partial charges on the atoms?

3. Express a wavenumber of 2318 cm^{-1} in terms of frequency, wavelength and energy.

4. Lambert's Law states that each successive layer of thickness dx of a medium absorbs an equal fraction

$$-\frac{dI}{I}$$

of radiation of intensity I, so that

$$-\frac{dI}{I} = b\,dx$$

where b is a constant. Obtain an expression for the intensity I of light which has travelled through a distance l, assuming that $I = I_o$ when $l = 0$.

5. Light of wavelength 257 nm is passed through a 3.00 cm cell filled with a solution of phenylalanine. What is the concentration of the solution if the absorbance is recorded as 0.81 and the absorption coefficient is 8850 dm^3 mol^{-1} cm^{-1}?

6. (a) The rotational energy levels $F(J)$ of a non-rigid molecule are given by the formula

$$F(J) = BJ(J + 1) - DJ^2(J + 1)^2$$

where J is the rotational quantum number, B is the rotational constant and D is the centrifugal distortion constant. Obtain an expression for

$$F(J + 1) - F(J)$$

in terms of B and D.

(b) Determine the energy of the $J = 1$ to $J = 2$ transition in HCl, which has a rotational constant of 10.593 7 cm^{-1} and a centrifugal distortion constant of 4.1×10^{-5} cm^{-1}.

7. What is the rotation constant of AlH for which the reduced mass is $1.613\ 6 \times 10^{-27}$ kg and the bond length 165.9 pm? The moment of inertia can be calculated using the formula given on Worked Example 2.10 on page 49.

8. (a) The vibrational energy levels $G(v)$ of an anharmonic oscillator are given by the expression

$$G(v) = \left(v + \frac{1}{2}\right)\omega_e - \left(v + \frac{1}{2}\right)^2 \omega_e x_e$$

where v is the vibrational quantum number, ω_e is the vibrational frequency and x_e is the anharmonicity constant. Obtain an expression for $G(1) - G(0)$, which is the most likely vibrational transition since higher levels will only be very sparsely populated.

(b) Determine the wavelength of the $v = 0$ to $v = 1$ transition for HCl which has a fundamental vibrational frequency of $2\ 988.9$ cm^{-1} and an anharmonicity constant of 1.540×10^{-2}.

9. The wavenumbers $\tilde{v}$ of lines in the rotation-vibration spectrum of carbon monoxide form a series defined by

$$\frac{\tilde{v}(m)}{\text{cm}^{-1}} = 2\ 143.28 + 3.813\ m - 0.017\ 5\ m^2$$

where $m = ..., -2, -1, 0, 1, 2,....$

Obtain an expression for $\tilde{v}\ (m + 1) - \tilde{v}\ (m)$ and show that for positive values of m the line spacing decreases while for negative values it increases with. increasing magnitude of m.

10. The selection rule for rotational transitions in Raman spectroscopy is $\Delta J = 2$. Obtain an expression for the energy associated with this transition.

11. The peaks due to each group of protons in a molecule are split into $n + 1$ by an adjacent group of n protons, the intensities of these peaks being given by the coefficients of Pascal's Triangle. Describe the appearance of the nuclear magnetic resonance spectrum of isopropyl iodide, $(CH_3)_2CHI$.

12. The exponential decay function, exemplified by

$$f(t) = \begin{cases} 0 & \text{for } t < 0 \\ e^{-at} & \text{for } 0 < t < 1 \\ 0 & \text{for } t > 1 \end{cases}$$

is important in Fourier transform spectroscopy. Determine its Fourier transform $g(\omega)$.

Statistical Mechanics

9.1 INTRODUCTION

Thermodynamics only deals with a macroscopic picture of matter, and its principles have all been derived by observing experimental behaviour. In contrast, quantum mechanics is concerned with the microscopic view and it starts from a more theoretical point, even though many of its results do explain experimental behaviour. Invariably, in quantum mechanics at this level of treatment, we are considering only single atoms or molecules.

Statistical mechanics provides a link between these subjects. We start with the microscopic view of matter and apply statistical techniques to a large number of chemical entities in order to reproduce the macroscopic functions of thermodynamics. This requires a few mathematical techniques which we have not yet come across, as well as some of those met in earlier chapters.

9.2 MOLECULAR ENERGY DISTRIBUTIONS

Much of statistical mechanics considers the distribution of molecules across energy levels. As you might expect, when we are dealing with large numbers of molecules some of the ideas we need to use may not be immediately obvious. To overcome this problem, we can often learn much about the principles we need to apply by considering simple systems.

Worked Example 9.1

In how many ways can four molecules be distributed across three energy levels, such that two are in the first level and one is in each of the higher levels?

CHEMICAL BACKGROUND

Specifying the distribution of molecules in this way is equivalent to specifying what is known as the state of the system. The more ways there are of achieving a given state, the more probable is the chance that the state will exist. In practice, it is found

DOI: 10.1201/9781003043218-9

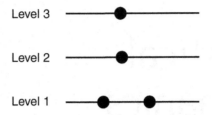

FIGURE 9.1 The distribution of molecules described in Worked Example 9.1.

that one state of system predominates, and this can aid the analysis considerably.

Solution to Worked Example 9.1

The arrangement of molecules within energy levels is illustrated in Figure 9.1. If we denote the four molecules as a, b, c, and d we can construct a table of possible arrangements as shown:

d	c	d	b	c	b	d	a	c	a	b	a
c	d	b	d	b	c	a	d	a	c	a	b
a, b	a, b	a, c	a, c	a, d	a, d	b, c	b, c	b, d	b, d	c, d	c, d

This can be constructed by first considering each pair of molecules which can occupy the lowest level, viz. a and b, a and c, a and d, b and c, b and d, and c and d. For each of these, the remaining two molecules can then occupy the upper levels in each of two ways. There are consequently twelve ways in which this can be achieved.

9.3 CONFIGURATIONS

The number of ways in which a specified number of particles can be arranged in each energy level is known as the number of configurations. For example, we might wish to specify two molecules in the lowest energy level, three in the next level and four in the level above. Such an arrangement can only be achieved in a certain number of ways starting with nine molecules.

Strictly speaking, we are dealing with instantaneous configurations, which will be constantly changing.

9.3.1 FACTORIALS

A factorial is a number written with an exclamation mark (!) immediately after it, such as 2!, 45! or 8! This strange-looking notation has a simple meaning, as illustrated by the following examples

$$5! = 5 \times 4 \times 3 \times 2 \times 1 = 120$$

$$9! = 9 \times 8 \times 7 \times 6 \times 5 \times 4 \times 3 \times 2 \times 1 = 362\,880$$

In general, for a number n, we have

$$n! = n \times (n-1) \times (n-2) \times (n-3) \dots 3 \times 2 \times 1$$

Worked Example 9.2

The number of configurations Ω is given by the equation

$$\Omega = \frac{N!}{\prod_i n_i!}$$

where N is the total number of particles and n_i is the number of particles in energy level i. The symbol $\prod$ denotes that we calculate the product of the quantities specified. What is the number of configurations if 3 molecules are in level 1, 2 molecules in level 2 and 1 molecule in level 3, as shown in Figure 9.2?

CHEMICAL BACKGROUND

In macroscopic terms, the arrangement of molecules between these energy levels is known as a state and will be constantly changing. Normally, the broadest distribution of the molecules across the energy levels will predominate. The expression for the number of configurations provides a starting point for the derivation of the equation which describes the Boltzmann distribution law.

Solution to Worked Example 9.2

The first stage is to write out the quantities which we have been given in the question using the notation of the given equation. If there are n_i molecules in level i, we can write

$$n_1 = 3,\ n_2 = 2,\ n_3 = 1$$

and the total number N of molecules is then given as

$$N = n_1 + n_2 + n_3$$

$$= 3 + 2 + 1$$

$$= 6$$

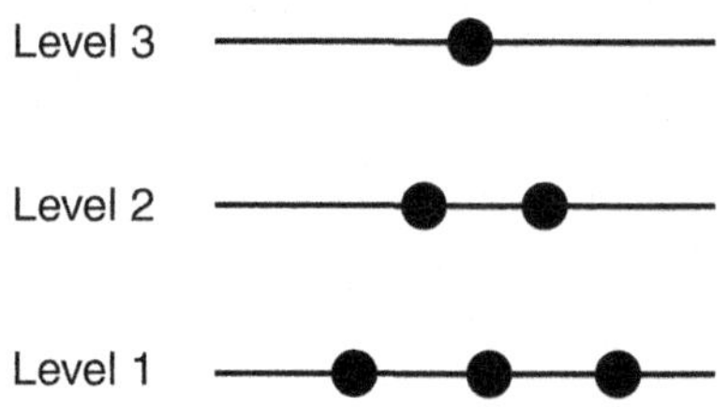

FIGURE 9.2 The distribution of molecules described in Worked Example 9.2.

We can now calculate $N!$ as

$$N! = 6!$$

$$= 6 \times 5 \times 4 \times 3 \times 2 \times 1$$

$$= 720$$

The other factorials required are

$$n_1! = 3! = 3 \times 2 \times 1 = 6$$

$$n_2! = 2! = 2 \times 1 = 2$$

and

$$n_3! = 1! = 1$$

We can now substitute into our equation to give

$$\Omega = \frac{N!}{\prod_i n_i!}$$

$$= \frac{N!}{n_1! n_2! n_3!}$$

$$= \frac{6!}{3! 2! 1!}$$

$$= \frac{720}{6 \times 2 \times 1}$$

$$= \frac{720}{12}$$

$$= 60$$

There is, however, a slightly neater way of performing this calculation. It will be shown here since the technique can be of use when we are dealing with symbols rather than actual numbers.

Since

$$6! = 6 \times 5 \times 4 \times 3 \times 2 \times 1$$

and

$$3! = 3 \times 2 \times 1$$

we can write

$$6! = 6 \times 5 \times 4 \times 3!$$

If we now substitute into the equation for Ω, we obtain

$$\Omega = \frac{6!}{3!\,2!\,1!}$$

$$= \frac{6 \times 5 \times 4 \times 3!}{3!\,2!\,1!}$$

Cancelling 3! top and bottom, and writing 2! as 2, and 1! as 1 gives

$$\Omega = \frac{6 \times 5 \times 4}{2}$$

$$= 6 \times 5 \times 2$$

$$= 60$$

This second method makes the calculation easier when a calculator is not available.

9.4 THE BOLTZMANN EQUATION

The Boltzmann distribution law allows us to specify the distribution of molecules among the energy levels. An increase in temperature results in an increase in molecular energy, as shown in Figure 9.3. Therefore, we would expect such a function to depend upon temperature.

The essential process in its determination is to maximize - the number of configurations Ω with respect to the number of particles in each level n_i. If Ω is a maximum, it follows that $\ln \Omega$ will also be a maximum, so that an alternative derivative which can be calculated and set equal to zero is

$$\frac{d \ln \Omega}{dn_i}$$

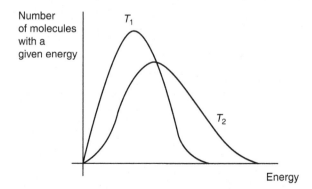

FIGURE 9.3   The Boltzmann distribution with $T_2 > T_1$.

9.4.1 DIFFERENTIATION OF LOGARITHMS

This topic was covered in Section 5.6.2 on page 178. The important result is

$$\frac{d\ln x}{dx} = \frac{1}{x}$$

9.4.2 DIFFERENTIATION OF PRODUCTS

This subject was described in Section 7.7.1 on page 263, where we saw that if $f(x) = u(x)\,v(x)$ then

$$\frac{df(x)}{dx} = v(x)\frac{du(x)}{dx} + u(x)\frac{dv(x)}{dx}$$

Worked Example 9.3

Stirling's approximation states that

$$\ln n! = n\ln n - n$$

for large values of n. Applying this to the derivation of the Boltzmann equation leads to the equation

$$\frac{d\ln\Omega}{dn_i} = -\frac{d}{dn_i}\left(n_i\ln n_i - n_i\right)$$

Determine $\dfrac{d\ln\Omega}{dn_i}$ by calculating the derivative on the right-hand side of this equation.

CHEMICAL BACKGROUND

Stirling's approximation is very useful when we are dealing with large numbers, as is the case in statistical mechanics. Consider how difficult it would be to calculate the factorial of Avogadro's number. If you try it on your calculator you may see that it cannot be done, but it is straightforward using the equation above.

Solution to Worked Example 9.3

The expression to be differentiated consists of two terms which will be considered in turn. The first involves determining

$$\frac{d}{dn_i}\left(n_i\ln n_i\right)$$

Remember that the rule for differentiating a product shown above is to differentiate the first function and multiply by the second, then differentiate the second function and multiply by the first. This will give us

$$\frac{d}{dn_i}\left(n_i \ln n_i\right) = \left(\ln n_i\right)\frac{d}{dn_i}\left(n_i\right) + n_i\frac{d}{dn_i}\left(\ln n_i\right)$$

Since

$$\frac{d}{dn_i}\left(n_i\right) = 1$$

and

$$\frac{d}{dn_i}\left(\ln n_i\right) = \frac{1}{n_i}$$

we obtain

$$\frac{d}{dn_i}\left(n_i \ln n_i\right) = \left(\ln n_i \times 1\right) + n_i \frac{1}{n_i}$$

$$= \ln n_i + 1$$

and so, since $\dfrac{d}{dn_i}\left(n_i\right) = 1$

$$\frac{d}{dn_i}\left(n_i \ln n_i - n_i\right) = -\left(\ln n_i + 1 - 1\right)$$

$$= -\ln n_i$$

Therefore

$$\frac{d\ln\Omega}{dn_i} = -\left(-\ln n_i\right) = \ln n_i$$

Worked Example 9.4

The Boltzmann equation for the number of particles n_i in an energy level i having energy ε_i at absolute temperature T is

$$\frac{n_i}{N} = \frac{\exp\left(-\dfrac{\varepsilon_i}{kT}\right)}{\displaystyle\sum_i \exp\left(-\dfrac{\varepsilon_i}{kT}\right)}$$

where N is the total number of particles and k is the Boltzmann constant.

Obtain an expression for the number of particles in level i relative to that in level j.

CHEMICAL BACKGROUND

The quantity

$$\sum_i \exp\left(-\frac{\varepsilon_i}{kT}\right)$$

in this expression is known as the partition function and is usually given the symbol θ. This quantity is of great importance in statistical mechanics. We often calculate the partition functions for translation, vibration, rotation and electrons separately.

Solution to Worked Example 9.4

We can use the equation as given

$$\frac{n_i}{N} = \frac{\exp\left(-\dfrac{\varepsilon_i}{kT}\right)}{\sum_i \exp\left(-\dfrac{\varepsilon_i}{kT}\right)}$$

for level i, and rewrite it to give

$$\frac{n_j}{N} = \frac{\exp\left(-\dfrac{\varepsilon_j}{kT}\right)}{\sum_i \exp\left(-\dfrac{\varepsilon_i}{kT}\right)}$$

for level j. Notice that we are still considering energy levels i in the second expression; it really does not matter what symbol we use as long as that on the summation sign Σ, corresponds to that inside the summation (in this case on the energy level).

When we take the ratio of n_i and n_j, we obtain

$$\frac{n_i}{n_j} = \frac{\exp\left(-\dfrac{\varepsilon_i}{kT}\right)}{\exp\left(-\dfrac{\varepsilon_j}{kT}\right)}$$

since the denominator of the defining equations is the same for n_i and n_j as it applies to the same system. To simplify this expression, we need to use the fact that

$$\frac{a^x}{a^y} = a^{x-y}$$

as shown in Section 1.6.2 on page 17, so we obtain

$$\frac{n_i}{n_j} = \exp\left(-\frac{\varepsilon_i}{kT} - \left(-\frac{\varepsilon_j}{kT}\right)\right)$$

We can rewrite

$$-\varepsilon_i - (-\varepsilon_j)$$

as

$$-\varepsilon_j + \varepsilon_j$$

which is given more usually as

$$- (\varepsilon_i - \varepsilon_j)$$

and so the expression we obtain is

$$\frac{n_i}{n_j} = \exp\left(\frac{-(\varepsilon_i - \varepsilon_j)}{kT} \right)$$

In the specific case where $i = 0$ and $j = 1$, we have

$$\frac{n_1}{n_0} = \exp\left(\frac{-\varepsilon_i}{kT} \right)$$

which can be written more succinctly as $\dfrac{n_1}{n_0} = e^{-\frac{\varepsilon_i}{kT}}$.

9.5 THE PARTITION FUNCTION

In the previous section, we met the idea of a partition function q, and saw that it was defined as

$$q = \sum_i \exp\left(-\frac{\varepsilon_i}{kT} \right)$$

This is actually the molecular partition function. The partition function Q for the whole system is equal to q^N, and this needs to be divided by $N!$ if the particles making up the system are indistinguishable, such as in a gas.

9.5.1 INTEGRATION BY SUBSTITUTION

In Section 7.7.2 on page 265, we met the technique of integration by parts, which was used for the integration of certain types of functions which could not be treated using the simple rules we had already learned. Other functions may be integrated by making an appropriate substitution.

Suppose we need to calculate the integral

$$\int_0^1 (x+5)^4 dx$$

One way of doing this would be to expand the bracket using Pascal's Triangle as outlined in Section 8.9.1 on page 307 and then to integrate each term of the polynomial individually. However, a quicker method which is of more general use is as follows.

If we make the substitution

$$u = x + 5$$

we can differentiate this defining equation to give

$$\frac{du}{dx} = 1$$

since the constant 5 will differentiate to zero. We can then rearrange the expression for the derivative to give

$$dx = du$$

Since this is a definite integral, we can replace our limits in the variable x with the corresponding limits in the variable u. For the lower limit, when $x = 0$, we have

$$u = x + 5$$

$$= 0 + 5$$

$$= 5$$

For the upper limit, $x = 1$ and so

$$u = x + 5$$

$$= 1 + 5$$

$$= 6$$

We are now able to work through the original expression for the integral, replacing terms relating to x with those relating to u. This involves:

- replacing the lower limit $x = 0$ with $u = 5$
- replacing the upper limit $x = 1$ with $u = 6$
- replacing $(x + 5)^4$ with u^4
- replacing dx with du.

The result of these replacements is that the integral to be calculated becomes

$$\int_5^6 u^4 \, du$$

This is straightforward to calculate using the basic rule of raising the power by one and dividing it by the new power. We then have

$$\int_5^6 u^4 du = \left[\frac{u^5}{5}\right]_5^6$$

Substituting the appropriate limits and using a calculator to evaluate the terms then gives

$$\int_5^6 u^4 du = \frac{6^5}{5} - \frac{5^5}{5}$$

$$= \frac{7776 - 3125}{5}$$

$$= \frac{4651}{5}$$

$$= 930.2$$

Worked Example 9.5

The rotational partition function q_r of a diatomic molecule is given in terms of the rotational quantum number J as

$$q_r = \int_0^\infty (2J+1)\exp\left(-\frac{J(J+1)h^2}{8\pi^2 IkT}\right) dJ$$

where I is the moment of inertia, h is Planck's constant, k is Boltzmann's constant and T is the absolute temperature. The upper limit on the integral sign is infinity, denoted by ∞. Calculate this integral in terms of these quantities by using the substitution $x = J(J+1)$.

CHEMICAL BACKGROUND

The fact that the rotational partition function is calculated using an integral rather than a summation is due to the rotational levels being closely spaced. We are effectively treating the rotational levels as being continuous rather than discretely spaced, and under these conditions, it is appropriate to use the integral.

The actual expression for q_r for diatomic molecules contains an additional quantity or known as the symmetry number. This takes the value 1 for heteronuclear diatomics (such as HCl) and 2 for homonuclear diatomics (such as Cl_2) to allow for the fact that in the latter a rotation of $180°$ produces an indistinguishable orientation.

Solution to Worked Example 9.5

The substitution suggested in the question is clearly related to the exponential term in the expression we want to integrate. As a first step, it is useful to remove the brackets so we have

$$x = J(J+1) = J^2+J$$

This can now be differentiated term by term, using the rule that we multiply by the power and reduce the power by one. This gives

$$\frac{dx}{dJ} = 2J+1$$

so that

$$dx = (2J+1)dJ$$

and

$$dJ = \frac{dx}{2J+1}$$

We can now separate the variables by multiplying each side by dJ and dividing by $2J+1$ to give

$$dJ = \frac{dx}{2J+1}$$

It is straightforward to change the lower limit in J to one in x. When $J = 0$, we have

$$x = 0^2 + 0 = 0$$

Similarly, when $J = \infty$ it follows that $x = \infty$. It is not really correct to use infinity as a number in the above equation, but it should be apparent that the two limits are the same.

We can now substitute for each term in the original expression. It is actually easier to leave the initial $(2J + 1)$ bracket as it is, but the other changes required are:

- to replace $J(J + 1)$ by x in the exponential term
- to replace dJ by $\dfrac{dx}{2J+1}$

It so happens that, in this example, the limits on the integral sign remain unchanged. The integral to be calculated in terms of x is therefore

$$q_r = \int_0^\infty (2J+1)\exp\left(-\frac{xh^2}{8\pi^2 IkT}\right)\frac{dx}{(2J+1)}$$

The terms in $2J+1$ cancel top and bottom and we are left with

$$q_r = \int_0^\infty \exp\left(-\frac{xh^2}{8\pi^2 IkT}\right) dx$$

This expression looks more complicated than it really is, and it is actually quite straightforward to integrate. The only variable in the exponential term is x, so we can group the other quantities to form a constant A defined by

$$A = \frac{h^2}{8\pi^2 IkT}$$

We can now write a much simpler expression for the integral to be calculated, which is

$$q_r = \int_0^\infty e^{-Ax} dx$$

remembering that e^{-Ax} is exactly the same $\exp(-Ax)$. We saw in Section 7.7.3 on page 266 that when we integrate the exponential function we obtain the same function divided by any constants. In this case, the constant is $-A$, so we obtain

$$q_r = \int_0^\infty e^{-Ax} dx$$

$$= \left[\frac{e^{-Ax}}{-A}\right]_0^\infty$$

Instead of dividing by $-A$ within the brackets, this term can be brought outside to give

$$q_r = -\frac{1}{A}\left[e^{-Ax}\right]_0^\infty$$

Before we actually evaluate this expression at the specified limits, it is worth realizing that since

$$a^{-n} = \frac{1}{a^n}$$

the expression for q_r can be written as

$$q_r = -\frac{1}{A}\left[\frac{1}{e^{Ax}}\right]_0^\infty$$

If we now consider what happens at the upper limit of infinity, we see that we are taking the reciprocal of the exponential of a very large number. The exponential of a very large number is itself very large, so the reciprocal of this will be very small and we will take it as zero. The lower limit is zero, so Ax will be zero and we need the reciprocal of e^0. Since any number raised to the power zero is one, the reciprocal of e^0 is also one and this is our lower limit. Substituting into the expression for q_r now gives us

$$q_r = -\frac{1}{A}[0-1]$$

$$= -\frac{1}{A}[-1]$$

The two negative signs multiply to give a positive sign, so we obtain

$$q_r = \frac{1}{A}$$

Since we chose to define A as

$$\frac{h^2}{8\pi^2 IkT}$$

we finally have

$$q_r = \frac{1}{A} = \frac{8\pi^2 IkT}{h^2}$$

9.5.2 CALCULATING A SERIES USING A SPREADSHEET

The concept of a sequence was introduced in Section 7.5.2 on page 252; if we sum the consecutive terms of a sequence we have a series. We also saw above that the partition function q can be defined as

$$q = \sum_i \exp\left(-\frac{\varepsilon_i}{kT}\right)$$

where ε_i are the energy levels in a molecule, k the Boltzmann constant and T the absolute temperature. Although we can obtain more usable expressions for the rotational and vibrational partition functions q_r and q_v, for the electronic partition function q_e we do usually sum over the actual energy levels. We then also have to take the degeneracy g_i into account, as explained in Section 8.6.1 on page 298, so the expression becomes

$$q = \sum_i g_i \exp\left(-\frac{\varepsilon_i}{kT}\right)$$

This type of repetitive calculation is ideally suited to a spreadsheet, where the numbers change in each term but not the form of the expression.

Worked Example 9.6

In the oxygen atom, the first five electronic energy levels are given in the table. Calculate the electronic partition function at 298 K using a spreadsheet.

Level	Energy/10^{-21} J	degeneracy
$^3P_2$	0	5
$^3P_1$	3.160	3
$^3P_0$	4.512	1
$^1D_2$	315.4	5
$^1S_0$	671.7	1

CHEMICAL BACKGROUND

In this example, each energy level is labelled according to its term symbol. This has the general form

$$^{2S+1}L_J$$

The quantum number L represents the total orbital angular momentum. When $L = 0$, we use the symbol S, when $L = 1$ the letter P and when $L = 2$ the letter D. The total angular momentum is represented by quantum number J, while S is the quantum number for spin angular momentum.

Solution to Worked Example 9.6

We begin by entering data for the energy and degeneracy into the spreadsheet. Put a label gi into cell B2 and ei into C2, with degeneracies in B4 to B8 and energies in C4 to C8. Note the way of entering a number in standard form, e.g. 3.160E-21 to represent 3.160×10^{-21}. These can be entered as given in the table above but will automatically be converted to standard form with one figure before the decimal point. We will also need values for k and T, so put these in cells F2 and F3 respectively, with labels in the preceding columns.

We now begin to perform the calculations by entering a formula in cell G4. This will be

$$= B4*EXP(-C4/(\$F\$2*\$F\$3))$$

Note the use of F2 and F3 to ensure that the values are always read from those actual cells, rather than using relative addressing. G4 should now contain the value 5.00E+00, since the exponent in this case is zero the value is simply 5.00.

Now copy the formula from G4 into the cells through to G8. Note that the values should decrease with increasing energy. The final stage is to sum all these terms. Go to cell G10, and enter the formula

$$= SUM(G4:G8)$$

	A	B	C	D	E	F	G
G10			f_x	=SUM(G4:G8)			
1							
2		gi	ei		k	1.38E-23	
3					T	298	
4		5		0			5.00E+00
5		3	3.16E-21				1.39E+00
6		1	4.51E-21				3.34E-01
7		5	3.15E-19				2.60E-33
8		1	6.72E-19				1.31E-71
9							
10						q	6.73E+00
11							

FIGURE 9.4 Spreadsheet used in Worked Example 9.6.

and the label q in cell F10. The shown value of 6.73E+00 is actually 6.73 as explained above.

It can actually be found by summing the first three terms only; a little inspection shows that the contribution higher levels make to the partition function is negligible. The final appearance of the spreadsheet is shown in Figure 9.4.

EXERCISES

1. If two six-sided dices are thrown, in how many ways can the numbers shown add to give an even number?

2. In how many ways could three cards belonging to the same suit of 13 cards be selected from a standard pack of playing cards?

3. Evaluate these quantities by cancelling as far as possible before using a calculator.

   a. $\dfrac{7!}{3!}$ b. $\dfrac{6!}{(5!)^2}$ c. $\dfrac{10!}{4! \times 2!}$ d. $\dfrac{6!\,4!}{5!\,3!}$

4. Determine the values of:

   a. $\dfrac{n!}{(n-1)!}$ b. $\dfrac{n!}{(n-4)!}$ c. $\dfrac{(n+2)!}{(n-1)!}$ d. $\dfrac{(n+1)!(n-1)!}{(n!)^2}$

5. Write each of these expressions in factorial form.

   a. $6 \times 5 \times 4$ b. $\dfrac{13 \times 12}{5 \times 4 \times 3 \times 2}$

   c. $\dfrac{9 \times 8 \times 7}{6 \times 5 \times 4 \times 3}$ d. $\dfrac{8 \times 7}{6 \times 5 \times 4 \times 3 \times 2}$

6. Write each of the following in factorial form:
   a. $n(n-1)(n-2)$
   b. $(n+1)n(n-1)$
   c. $(n+6)(n+5)(n+4)(n+3)$
   d. $(n+1)^2(n+2)$

7. Write each of these expressions as a multiple of a single factorial quantity.
   a. $6!+7!$
   b. $2(10!)+3(7!)$
   c. $10!-8!$
   d. $6(5!)-4(3!)$

8. Factorise the following expressions by writing them as a single factorial quantity:
   a. $n!-(n-1)!$
   b. $n!-(n-2)!$
   c. $(n+1)!+(n-1)!$
   d. $n!+(n+1)!+(n+2)!$

9. Show that

   $$\frac{d}{dx}(\ln kx) = \frac{1}{x}$$

   regardless of the value of k.

10. Obtain a general expression for

    $$\frac{d}{dx}(\ln \sqrt[m]{x^n})$$

11. Differentiate the functions:
    a. $x^2 \ln x$
    b. $x^3 e^{2x}$
    c. $x e^{-4x}$
    d. $x e^{2x}+x^2 \ln x$
    with respect to x.

12. Differentiate the following functions with respect to x:
    a. $x \sin x$
    b. $x^2 \cos x$
    c. $e^x \cos (3x+1)$
    d. $e^{-2x} \sin (4x-3)$

13. Determine these integrals by using an appropriate substitution:
    a. $\displaystyle\int_1^3 (x-4)^3 dx$

    b. $\displaystyle\int_0^1 (2x+1)^4 dx$

    c. $\displaystyle\int x\sqrt{x^2+2}\, dx$

    d. $\displaystyle\int_0^2 e^{2x+3}\, dx$

14. Determine the following integrals by using an appropriate substitution:

    a. $\displaystyle\int e^x \sqrt{e^x + 2}\, dx$ b. $\displaystyle\int_0^2 x(x^2 + 1)^3\, dx$

    c. $\displaystyle\int \frac{\ln x}{x}\, dx$ d. $\displaystyle\int_2^3 \frac{1}{x \ln x}\, dx$

15. Use an appropriate substitution to determine:

    a. $\displaystyle\int \frac{dx}{x-5}$ b. $\displaystyle\int_2^4 \frac{2}{3x+2}\, dx$ c. $\displaystyle\int \frac{5}{2x+1}\, dx$ d. $\displaystyle\int_6^8 \frac{3}{2x-5}\, dx$

16. Evaluate the following integrals by using an appropriate substitution:

    a. $\displaystyle\int_0^{\frac{\pi}{2}} \sin(3x+2)\, dx$ b. $\displaystyle\int \sin(x-5)\, dx$

    c. $\displaystyle\int_0^{\pi} \cos(2x-1)\, dx$ d. $\displaystyle\int_0^{\frac{\pi}{2}} \cos(5x+1)\, dx$

17. Use a spreadsheet to determine the sum of the series defined by

$$f(n) = n^2$$

from $n = 1$ to $n = 10$.

18. Use a spreadsheet to determine the value of

$$\sum_{n=1}^{12} \frac{1}{n^3}$$

19. Use a spreadsheet to evaluate the sum of the first ten terms of the series defined by

$$f(n) = 3^{n+2}$$

for $n \geq 1$.

20. Obtain a general expression for the series

$$2 + 9 + 16 + \dots + 107$$

and hence evaluate this sum using a spreadsheet.

PROBLEMS

1. From first principles, deduce how many ways there are of arranging five molecules in three energy levels so that there are two molecules in each of the two lowest levels.

2. The number of configurations Ω of a system is given by the expression

$$\Omega = \frac{N!}{\prod_i n_i!}$$

where N is the total number of particles and n_i is the number of particles in level i. The symbol $\prod$ denotes multiplication so $\prod_i n_i!$ means multiply all the values of $n_i!$ together. Obtain an expression for Ω when
   a. all N particles are in different energy levels, as shown in Figure 9.5.
   b. all particles except one are in the lowest level, as shown in Figure 9.6.

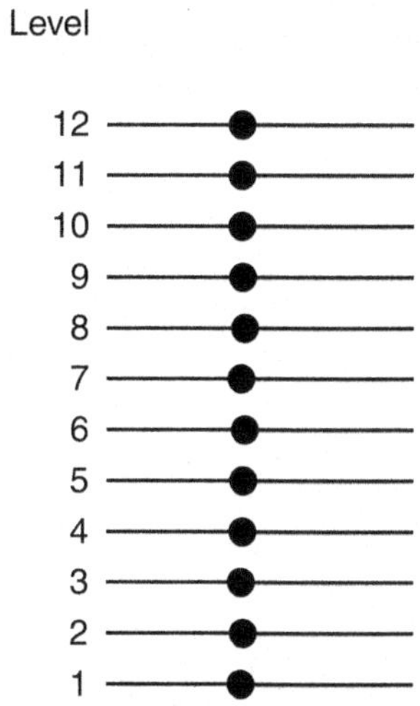

FIGURE 9.5 All particles in different energy levels.

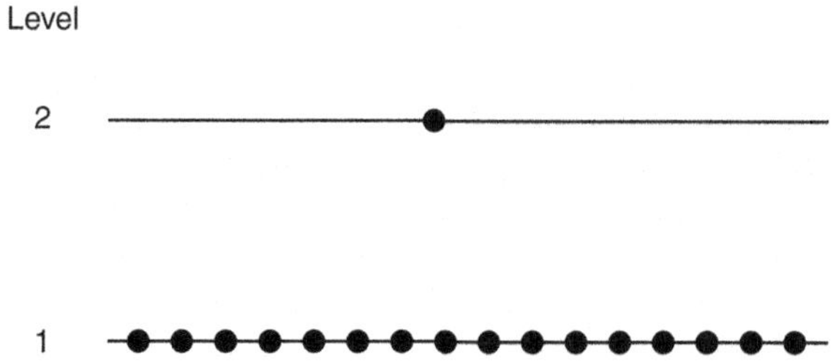

FIGURE 9.6 All particles except one in the lowest energy level.

3. The translational partition function q_{trans} for a monatomic gas is given by the expression

$$q_{trans} = \frac{(2\pi mkT)^{\frac{3}{2}}V}{h^3}$$

where m is the mass of one molecule, k is Boltzmann's constant, T is the absolute temperature, V is the enclosing volume and h is Planck's constant. Obtain an expression for the pressure p of the gas by using the formula

$$p = NkT\left(\frac{\partial \ln q_{trans}}{\partial V}\right)_T$$

where N is the number of molecules present.

4. The translational partition function q_t is defined by the equation

$$q_t = \frac{(2\pi mkT)^{\frac{3}{2}}V}{h^3}$$

where m is the mass of the atom or molecule at absolute temperature T in a volume V. Calculate the translational partition function for a hydrogen atom in a volume of $1.00\,m^3$ at 298 K.

5. (a) Obtain the equation for Stirling's approximation $\ln N! = N \ln N - N$ by writing $\ln N!$ as a sum from $n = 1$ to $n = N$, replacing it with an integral which can be evaluated using the result of question 20 in the Exercises in Chapter 7, and neglecting appropriate terms when N is large.

(b) Rewrite the expression for Stirling's approximation in terms of an exponential rather than a logarithmic function.

(c) What is the error introduced by using Stirling's approximation when $N = 10$?

6. The partition function q is defined by the equation

$$q = \sum_i g_i \exp\left(-\frac{\varepsilon_i}{kT}\right)$$

where ε_i is the energy of level i, k is Boltzmann's constant and T is the absolute temperature. For a system containing energy levels with the same spacing but a different zero, i.e $e_i + \alpha$ where α is a fixed constant, obtain an expression for the partition function q' in terms of q.

7. It is often helpful to express the energy levels of a molecule in terms of ε_i/k as this assists computation. The electronic energy levels for the nitrogen atom are expressed in this form below:

g_i	$\dfrac{\varepsilon_i}{k}$ /K
4	0
6	27 658.7
4	27 671.7
6	41 492.4

The degeneracy g_i of each level is also given. Use this data to show that only the ground state contributes to the electronic partition function of nitrogen at 298 K and hence calculate the value of the partition function.

8. The rotational constant B of a rigid rotor is defined by the equation

$$B = \frac{h^2}{8\pi^2 I}$$

where h is Planck's constant and I the moment of inertia.
a. The value of B for HCl is 10.593 7 cm^{-1}. Convert this wavenumber to an energy value.
b. Use the relationship

$$q_r = \frac{kT}{B}$$

to obtain the rotational partition function of HCl at 298.14 K. The absolute temperature in this equation is represented by T, and the Boltzmann constant by k.

9. The overall partition function is the product of the individual partition functions, and that due to electronic motion is generally taken to be 1. Calculate the partition function of Cl_2 for which the translational, rotational and vibrational partition functions are 3.52×10^{32}, 182 and 1.070 respectively.

10. The vibrational energy levels of HCl have an even spacing of 2990 cm^{-1}. Calculate the ratio of the number of molecules in one level to those in the next lowest level at (a) 25°C and (b) 100°C.

11. The statistical entropy S is given by the equation

$$S = k \ln \Omega$$

where Ω is the number of configurations and k is the Boltzmann constant. Calculate the statistical entropy associated with 1 mole of configurations.

12. The absolute entropy S is defined as the entropy change of a substance taken from absolute zero to a given temperature. For water at 298 K, this value is 69.95 J K^{-1}mol^{-1}. How many configurations does this represent?

Appendix A: Units

A.1 SI PREFIXES

Multiplier	Name	Symbol
10^{-18}	atto	a
10^{-15}	femto	f
10^{-12}	pico	p
10^{-9}	nano	n
10^{-6}	micro	μ
10^{-3}	milli	m
10^{-2}	centi	c
10^{-1}	deci	d
10^{3}	kilo	k
10^{6}	mega	M
10^{9}	giga	G
10^{12}	tera	T

A.2 EQUIVALENT UNITS

Unit	Symbol	Equivalent(s)	
Ångström	Å	10^{-10} m	
Hertz	Hz	s^{-1}	
Joule	J	N m	$kg\,m^2s^{-2}$
Litre	l	dm^3	$10^{-3}m^3$
Newton	N	$kg\,m\,s^{-2}$	
Pascal	Pa	$N\,m^{-2}$	$kg\,m^{-1}s^{-2}$
Volt	V	$J\,C^{-1}$	$kg\,m^2s^{-3}\,A^{-1}$

Appendix B: Physical Constants

These are given to four significant figures. There will be cases where you will need to find more precise values, and these will be indicated in the text.

Constant	Symbol	Value
Avogadro constant	L	$6.022 \times 10^{23}\,mol^{-1}$
Bohr radius	a_o	$5.292 \times 10^{-11}\,m$
Boltzmann constant	k	$1.381 \times 10^{-23}\,J\,K^{-1}$
Electronic charge	e	$1.602 \times 10^{-19}\,C$
Electron rest mass	m_e	$9.109 \times 10^{-31}\,kg$
Faraday constant	F	$9.649 \times 10^{4}\,C\,mol^{-1}$
Gas constant	R	$8.314\,J\,K^{-1}\,mol^{-1}$
Planck constant	h	$6.626 \times 10^{-34}\,J\,s$
Velocity of light in a vacuum	c	$2.998 \times 10^{8}\,m\,s^{-1}$

Answers to Exercises

CHAPTER 1

1. (a) 2 (b) −5 (c) −6 (d) 0
2. (a) −2 (b) 7 (c) −5 (d) 4
3. (a) 28 (b) −12 (c) 21 (d) −24
4. (a) 7 (b) −7 (c) 9 (d) −5
5. (a) 9 (b) −1 (c) 7 (d) −15
6. (a) $y = x - 10$ (b) $y = 8 - 3x$ (c) $y = 2x - 5$ (d) $y = \dfrac{2}{x}$
7. (a) 3^6 (b) x^6 (c) xy^5 (d) y^5
8. (a) 6^{-1} (b) 5^{-1} (c) x^{-1} (d) y^6
9. (a) 4^{12} (b) 3^{12} (c) x^{12} (d) y^{-12}
10. (a) $3^{\frac{5}{2}}$ (b) 6^{12} (c) $x^{\frac{5}{3}}$ (d) $y^{-\frac{3}{5}}$
11. (a) 4^{-3} (b) 8^{-6} (c) $3x^{-5}$ (d) $2y^2$
12. (a) $\dfrac{4}{5}$ (b) $\dfrac{2}{3}$ (c) $\dfrac{4}{5x}$ (d) $\dfrac{2x}{3y}$
13. (a) $\dfrac{37}{28}$ (b) $\dfrac{13}{8}$ (c) $\dfrac{1}{9}$ (d) $\dfrac{1}{6}$
14. (a) $\dfrac{21}{32}$ (b) $\dfrac{16}{27}$ (c) $\dfrac{5}{21}$ (d) $\dfrac{5}{16}$
15. (a) $\dfrac{9}{10}$ (b) $\dfrac{3}{4}$ (c) $\dfrac{9}{10}$ (d) $\dfrac{40}{27}$
16. (a) $3.638\,7 \times 10^4$ (b) $2.238\,000 \times 10^6$ (c) 1.76×10^{-4} (d) 1.526×10^{-2}
17. (a) 36 200 000 (b) 16 600 (c) 0.000 012 7 (d) 0.003 948
18. (a) $-\dfrac{59}{36}$ (b) $-\dfrac{3}{16}$ (c) $\dfrac{127}{72}$ (d) $-\dfrac{8}{9}$
19. (a) −1 (b) 32 (c) −117 (d) 102
20. (a) $\dfrac{5}{24}$ (b) $-\dfrac{67}{48}$ (c) $\dfrac{41}{72}$ (d) $-\dfrac{57}{40}$

CHAPTER 2

1. (a) 350 (b) 0.001 2 (c) 0.54 (d) 13 (e) 4300
2. (a) 13.85 (b) 0.25 (c) 99.54 (d) 0.01 (e) 1 200.00
3. (a) 78 500 (b) 0.006 75 (c) 12.0 (d) 80 000 (e) 0.319
4. (a) 3.1 (b) 12.8 (c) 0.0 (d) 1.0 (e) 1.0
5. (a) 15.6 (b) 19.1 (c) 2.3 (d) 2.2 (e) 13.9 (f) 13.3
6. (a) 65.48 (b) 1160 (c) 1.24 (d) 2.3

7. (a) 10.4 (b) 4.4 (c) 1.72 (d) 8.55
8. (a) 2.85 (b) 0.36 (c) 3.62 (d) 0.321
9. (a) 0.4 (b) 0.01 (c) 1%
10. (a) 186 cm^3 (b) 0.017 (c) 1.7%
11. (a) 108 cm^3 (b) 0.01 (c) 1%
12. (a) 0.14 (b) 0.34 (c) 0.10 (d) 0.31
13. (a) 0.006 (b) 0.01 (c) 0.4 (d) 0.003
14. 31.4 ± 0.2 m s^{-1}
15. (a) 10.63 (b) 10.62 (c) 10.61
16. Average = 23.89, $n = 6$, $\Sigma(x_i - \bar{x})^2 = 0.138$, variance = 0.028, standard deviation = 0.17.
17. $n = 7$, average = 99.56, $\Sigma(x_i - \bar{x})^2 = 4.420$, variance = 0.631, standard deviation = 0.79, standard error = 0.30.
18. 0.08
19. Standard deviation = 1.0, standard error = 0.4.
20. 0.04

CHAPTER 3

1. (a) $y = 7 - x$ (b) $y = 3x - 6$ (c) $y = \dfrac{12}{x}$ (d) $y = \dfrac{x}{12}$

2. (a) $y = \dfrac{x}{1-x}$ (b) $y = \dfrac{2}{\sqrt{x}}$ (c) $y = \dfrac{5}{x}$ (d) $y = \dfrac{x^2}{10x^2 - 2}$

3. (a) $x^3y + xy^3$ (b) $x^3 + x^2y - x^2y^2 - xy^3$ (c) $x^2 - y^2$ (d) $x^3y - bx^3 + ax^2y - abx^2$

4. (a) (i) 4 (ii) 228 (iii) $12x^3 + 6x^2 + 8x + 1$
   (b) (i) 6 (ii) 5589 (iii) $48x^5 + 5x^4$
   (c) (i) 3 (ii) –215 (iii) $24x^2$
   (d) (i) 2 (ii) 32 (iii) $6x - 2$

5. $-\dfrac{1}{x^2} + \dfrac{4}{x^3} - \dfrac{9}{x^4} + \dfrac{16}{x^5} - \dfrac{25}{x^6}$, -8.145

6. (a) 1.380 (b) 1 (c) 1.099 (d) 1.000
7. (a) gradient = 6, intercept = 3
   (b) gradient = $\dfrac{5}{3}$, intercept = $\dfrac{4}{3}$
   (c) gradient = $\dfrac{1}{2}$, intercept = -4
   (d) gradient = 4, intercept = 0

8.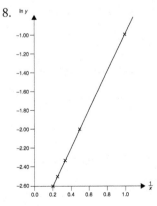

Gradient = 2.0.

9.
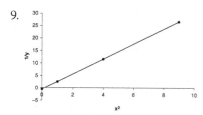

Gradient = 2.9.

10. 3

11. 1.8

12. Inversely proportional with a proportionality constant of 64.

13. (a) −1 (b) 0 (c) 16 (d) −32

14. (a) 0 (b) 0 (c) $\dfrac{3}{2}$ (d) $-\dfrac{3}{2}$

   $f(x, y)$ is not defined when $x = 0$ or $y = 0$.

15. (a) 1 (b) 2.69 (c) 4.10 (d) −0.61

16. $\dfrac{\partial f}{\partial x} = 6x + 2y + 12x^2 y^2, \dfrac{\partial f}{\partial y} = 2x + 8x^3 y$

   $df(x, y) = (6x + 2y + 12x^2 y^2)dx + (2x + 8x^3 y)dy$

17. $\dfrac{\partial^2 f}{\partial x \partial y} = \dfrac{\partial^2 f}{\partial y \partial x} = 2 + 24x^2 y$

18. $2x + 2xy + y^2, x^2 + 2xy + 2y$

19. (a) $x = 0.72$ or $x = -9.72$

   (b) $x = -1.39$ or $x = -0.36$

   (c) $x = 0.24$ or $x = 4.27$

   (d) $x = 5.97$ or $x = -1.57$

20. $x = -1.3$ or $x = 0.5$

CHAPTER 4

1. (a) 0.42 (b) 0.81 (c) 0.2 (d) 1.206

2. (a) −1.523 (b) 1.319 (c) 0.504 (d) 1.085

3. (a) 1.03 (b) 2.39 (c) 3.00 (d) 3.81

4. (a) 3 (b) 8 (c) 3 (d) 2

5. (a) 16 (b) 9 (c) 11.18 (d) 32.0

6. (a) x is inversely proportional to y.

   (b) x^2 is inversely proportional to y.

   (c) x is directly proportional to y.

   (d) x^2 is directly proportional to y.

7. y is directly proportional to x^2.

8. (a) $r = \dfrac{Kxy}{z^2}$ (b) $r = \dfrac{Kxz^2}{y}$ (c) $r = \dfrac{Kyz^2}{x}$ (d) $r = \dfrac{Ky}{xz^2}$

   K = constant of proportionality

9. (a) y against ln x, gradient = 6, intercept = 4

   (b) log y against $\sqrt{x}$, gradient = 2, intercept = 0

   (c) ln y against log x, gradient = 2, intercept = −3

   (d) $1/y$ against x^2, gradient = 1, intercept = −7

10. (a) t (on y-axis) against s^2 (on x-axis), gradient = 1, intercept = −5
    (b) w (on y-axis) against v (on x-axis), gradient = 3, intercept = 9
    (c) a (on y-axis) against b (on x-axis), gradient = 1, intercept = 14
       If other way around gradient = 1, intercept = −14
    (d) c^2 (on y-axis) against d (on x-axis), gradient = −1, intercept = 64
       If other way around gradient = −1, intercept = 64

11. (a) y against $\dfrac{1}{x}$, gradient = 4, intercept = 7

    (b) $\log 2y$ against $\dfrac{1}{x^2}$, gradient = 1, intercept = 6

    (c) $\dfrac{1}{y^2}$ against $\dfrac{1}{x}$, gradient = 1, intercept = ½

    (d) y^2 against $\dfrac{1}{x^2}$, gradient = 2, intercept = −¾

12. $y = 1.8x − 0.4$
13. $a = 4.2$, $b = −3.8$
14. (a) $f(1) = 11$, (b) $f(−1) = 5$, (c) $f(−2) = 14.75$, (d) $f(0.864) = 9.105$
15. (a) $x = −3$, $x = 3$ (b) $x = −1.5$ (c) $x = 0$ (d) $x = −1$, $x = 1$
16. (a) $x = 0$ and $y = 0$ (b) $x = 0$ or $x = 1$ (c) $y = −1$ or $y = 1$ (d) $y = 4$
17. (a) $x = 1$ and $y = 0$ (b) $x = −1$, $x = 1$ and $y = 0$
    (c) $x = 1$, $y = −1$ and $y = 1$ (d) $x = −1$, $x = 1$, $y = 0$
18. (a) $\dfrac{df(x)}{dx} = 45x^4 + 9x^2 − 2$, $\dfrac{d^2f(x)}{dx^2} = 180x^3 + 18x$

    (b) $\dfrac{df(x)}{dx} = 32x^3 − 4x$, $\dfrac{d^2f(x)}{dx^2} = 96x^2 − 4$

    (c) $\dfrac{dg(y)}{dy} = 4y + \dfrac{3}{y^2}$, $\dfrac{d^2g(y)}{dy^2} = 4 − \dfrac{6}{y^3}$

    (d) $\dfrac{dg(y)}{dy} = −\dfrac{1}{y^2} + \dfrac{4}{y^3} − \dfrac{9}{y^4}$, $\dfrac{d^2g(y)}{dy^2} = \dfrac{2}{y^3} − \dfrac{12}{y^4} + \dfrac{36}{y^5}$

19. (a) $x = −2$, $x = 2$ (b) $x = ⅓$ (c) $x = 0.395$ (d) $x = 1$
20. (a) Minimum at $x = 7/6 = 1.17$.
    (b) Minimum at $x = −3/8 = −0.375$.
    (c) Point of inflexion at $x = 0$.
    (d) Maximum at $x = ¼ = 0.25$.

CHAPTER 5

1. 1.9

2. (a) $\dfrac{x^6}{6} + C$ (b) $x^7 + C$ (c) $−\dfrac{1}{3x^3} + C$ (d) $−\dfrac{2}{x} + C$

3. (a) $2x^3 + \dfrac{9}{2}x^2 + 8x + C$ (b) $\dfrac{3}{4}x^4 + \dfrac{4}{3}x^3 + C$

   (c) $\dfrac{3}{5}x^5 − \dfrac{4}{3}x^3 + 6x + C$ (d) $\dfrac{x^6}{3} − x^4 + \dfrac{x^2}{2} + 7x + C$

4. (a) $\dfrac{2}{3}$ (b) 48 (c) 0.67 (d) 28 068

5. (a) $\frac{1}{2}x^2 + \ln x + C$ (b) 6.746 (c) 2 425.8 (d) $\frac{5}{6}$

6. (a) $-\frac{1}{2x^2} + C$ (b) 0.799 7 (c) -6.72×10^{-5} (d) $\frac{1}{6}$

7. (a) $x^3 + 5\ln x - \frac{1}{x^2} + C$ (b) 666 620.8

(c) $\frac{9}{8}x^8 - \frac{4}{3}\ln(3x-1) + \frac{3}{5x^5} + C$ (d) 40.049

8. (a) log 12 (b) ln 6 (c) ln 2 (d) log 1296

9. $\frac{x}{(x+2)(x+3)} = \left(\frac{3}{x+3}\right) - \left(\frac{2}{x+2}\right)$

10. (a) $4x + \frac{3}{x}$ (b) $3 + \left(\frac{2x}{x^2-2}\right)$ (c) $12x^2 - 4x\ln(2x^2+1)$ (d) $\frac{4}{(4x+3)} + \frac{3}{(3x-2)}$

11. (a) $\frac{x^2}{2} + \ln(x-2) + C$ (b) 0.444 (c) -1.262 (d) 0.175

12. $-0.052\ 2$

13. (a) 0.099 6 (b) 29.56 (c) 3 (d) 0.306

14. (a) e^6 (b) e^9 (c) x^2e^3 (d) x^2e

15. (a) $x = 1.623$ (b) $x = -0.288$ (c) $x = 1.082$ (d) $x = 6.374$

16. (a) $x = 36.6$ (b) $x = 335$ (c) $x = 62\ 946$ (d) $x = 2.061$

17. (a) $arcf(x) = \sqrt{\frac{x-5}{3}}$ (b) $arcg(x) = \left(\frac{x+8}{2}\right)^2$

(c) $arcF(x) = \left[\frac{7}{8}(x-2)\right]^2$ (d) $arcG(x) = \sqrt{\frac{(e^x - 2)}{3}}$

18. $x = 6.5, y = -10.5$

19. $x = -2, y = 3$

20. $x = 2, y = -2, z = 0$

CHAPTER 6

1. 9.51 cm

2. (a) 2.8 (b) 5.0 (c) 7.6 (d) 2.8

3. (a) 3.0 (b) 2.5 (c) 9.5 (d) 3.7

4. $\sin\theta = 0.82$, $\cos\theta = 0.59$, $\tan\theta = 1.38$

5. 34.9°, 55.1°, 90.0°

6. (a) 0.191 (b) 0.719 (c) 0.848 (d) 0.414

7. (a) 0.866 0 (b) $-1.000\ 0$ (c) $-0.707\ 1$ (d) 0.577 4

8. (a) $x = 20.6°$ (b) $x = 66.0°$ (c) $x = 96.2°$ (d) $x = 37.4°$

9. (a) $x = -0.280$ rad (b) $x = -0.343$ rad (c) $x = 1.585$ rad (d) -1.107 rad

10. (a) 4.1 (b) 7.5 (c) 6.2 (d) 4.6

11. (a) $0.53\ \mathbf{i} - 0.27\mathbf{j} + 0.80\mathbf{k}$

(b) $-0.14\mathbf{i} - 0.27\mathbf{j} + 0.95\mathbf{k}$

(c) $0.30\mathbf{i} - 0.90\mathbf{j} + 0.30\mathbf{k}$

(d) $-0.53\mathbf{i} + 0.80\mathbf{j} - 0.26\mathbf{k}$

12. (a) $3\mathbf{i} + \mathbf{j} - 2\mathbf{k}$ (b) $-\mathbf{i} + 3\mathbf{j} - 4\mathbf{k}$

(c) -3 (d) $-\mathbf{i} - 7\mathbf{j} - 5\mathbf{k}$

13. 105.8°
14. (a) real 4, imaginary 2 (b) real 1, imaginary -8
    (c) real 3, imaginary -2 (d) real-4, imaginary -7
15. (a) $7 + 3i$ (b) $8 + 2i$ (c) $2-4i$ (d) $5 + 6i$
16. (a) $17 + i$ (b) $-1 + 5i$ (c) $24 - 10i$ (d) $9 + 20i$
17. $\cos kx = \dfrac{e^{ikx} + e^{-ikx}}{2}$, $\sin kx = \dfrac{e^{ikx} - e^{-ikx}}{2i}$

18. (a) $\begin{pmatrix} 3 & 2 \\ 7 & 3 \end{pmatrix}$ (b) $\begin{pmatrix} 1 & 2 \\ -1 & -1 \end{pmatrix}$ (c) $\begin{pmatrix} 1 & 0 \\ -6 & -4 \end{pmatrix}$ (d) $\begin{pmatrix} 10 & 4 \\ 7 & 2 \end{pmatrix}$

19. $\begin{pmatrix} 7 \\ 10 \\ 6 \end{pmatrix}$

20. $\begin{pmatrix} 32 & 22 \\ 31 & 32 \end{pmatrix}$

CHAPTER 7

1. (a) 9.006 53 (b) 8.204 707 47 (c) 7.666 854 (d) 17.760 919 33
2. $xy = 10.11$
3. (a) 611 (b) 0.007 35
4. Minimum at $x = -0.135$.
5. Minimum in $f(x)$ at (1,6), minimum in $g(x)$ at (1,0).
6. 5
7. 12.76
8. (a) $4-6i$ (b) $-2 + i$ (c) $9 + 4i$ (d) $-3 - 2i$
9. $6\left(\cos\dfrac{7\pi}{12} - i\sin\dfrac{7\pi}{12} \right)$
10. 2, 12, 30, 56, 90
11. (a) 16, 19 (b) 37, 50 (c) 26, 51 (d) 49, 71
12. $f(n) = 3 \times 2^n$ for $n \geq 0$
13. (a) $arcf(x) = \dfrac{\sqrt{x}-1}{2}$ (b) $arcg(x) = \dfrac{\sqrt[3]{x}+1}{3}$ (c) $arch(x) = \sqrt{\sqrt{x} - 3}$
14. (a) $arcf(x) = e^x - 2$ (b) $arcg(x) = \dfrac{\cos^{-1}x - 1}{4}$ (c) $arch(x) = \sqrt{\dfrac{1}{a}\ln\left(\dfrac{x}{2}\right)}$
15. (a) $0.3x^{-0.7}$ (b) $-1.2x^{-1.6}$ (c) $6y^{-2.5}$ (d) $-t^{-1.5}$
16. (a) $2e^{2x}$ (b) $-3e^{-3x}$ (c) $3\exp(3x+2)$ (d) $2\exp(2x-1)$
17. (a) $2x \ln x + x$ (b) $e^{2x}(2x+1)$ (c) $2y\,e^{-4y} - 4y^2\,e^{-4y}$ (d) $\sin 3y + 3y\cos y$
18. (a) $\dfrac{e^{4x}}{4} + C$ (b) 402.93 (c) -40.12 (d) 51.55
19. $\dfrac{(e^2 + 1)}{4} = 2.09$
20. $x \ln x - x + C$
21. (a) -33 (b) -1 (c) -2 (d) 8
22. (a) 7 (b) 13 (c) 7 (d) 0

CHAPTER 8

1. 6.6, 5.3
2. 0.380
3. −4.1
4. (a) x is inversely proportional to $\ln y$
   (b) x^2 is directly proportional to y
   (c) x is inversely proportional to e^{-y}
   (d) x is directly proportional to e^y

5. (a) y is directly proportional to x, constant $= \dfrac{a}{b}$

   (b) y is inversely proportional to x^2, constant $= a^2b$

   (c) y is directly proportional to $\ln x$, constant $= \dfrac{a}{b}$

   (d) y is inversely proportional to e^x, constant $= \dfrac{a}{b}$

6. (a) $\ln x$ (b) $\ln x^3 y$ (c) $\ln\left(\dfrac{1}{y^2}\right)$ (d) $\ln x - \ln y$

7. (a) $\tfrac{1}{2}\ln (x + 1) - \tfrac{1}{2}\ln (x - 1)$
   (b) $\ln 5 + 4\ln x$
   (c) $\ln 3 - 2\ln x + \ln y - \ln z$

   (d) $\ln x + \ln y - \dfrac{1}{2}\ln z$

8. (a) $\log 2$ (b) $\log x$ (c) $\log (5x^2)$ (d) $\log (x^2 - 1)$
9. 31
10. $2N + 3$
11. 2
12. (a) 0 (b) $\cos x = b + 2cx$ (c) 1 (d) 0 (e) $\sin x = x$
13. (a) $x^4 + 12 x^3 + 54 x^2 + 108 x + 81$
    (b) $16x^4 - 96x^3 + 216x^2 - 216x + 81$
    (c) $32 x^5 - 320 x^4 + 1280 x^3 - 2560 x^2 + 2560 x - 1024$
    (d) $729 x^6 + 2916 x^5 + 4860 x^4 + 4320 x^3 + 2160 x^2 + 576 x + 64$
14. (a) $x^5 + 5x^4 y + 10x^3 y^2 + 10x^2 y^3 + 5xy^4 + y^5$

    (b) $\dfrac{81}{x^4} - \dfrac{108y}{x^3} + \dfrac{54y^2}{x^2} - \dfrac{12y^3}{x} + y^4$

    (c) $16x^4 + 160x^3 y + 3000x^2 y^2 + 1000xy^3 + 625y^4$

    (d) $\dfrac{32}{x^5} + \dfrac{240}{x^4 y} + \dfrac{720}{x^3 y^2} + \dfrac{1080}{x^2 y^3} + \dfrac{810}{xy^4} + \dfrac{243}{y^5}$

15. −108
16. 0.988 060
17. $f(n) = 2^{n-1}$
18. Write Pascal's triangle in form
    1
    1 1
    1 2 1
    1 3 3 1
    1 4 6 4 1
    Diagonals are then 1, 1 + 1, 2 + 1, 1 + 3 + 1, 1 + 4 + 3 etc.

19. $\dfrac{-3i\sin\pi y}{\pi(9-y^2)}$

20. $-\dfrac{1}{2\pi iy}e^{-iy}+\dfrac{1}{\pi y^2}e^{-iy}+\dfrac{1}{\pi iy^3}e^{-iy}-\dfrac{1}{\pi iy^3}$

CHAPTER 9

1. 18
2. 1716
3. (a) 840 (b) 0.05 (c) 75 600 (d) 24
4. (a) n (b) $n(n-1)(n-2)(n-3)$ (c) $(n+2)(n+1)n$ (d) $\dfrac{n+1}{n}$
5. (a) $\dfrac{6!}{3!}$ (b) $\dfrac{13!}{5!\,11!}$ (c) $\dfrac{2!9!}{(6!)^2}$ (d) $\dfrac{8!}{(6!)^2}$
6. (a) $\dfrac{n!}{(n-3)!}$ (b) $\dfrac{(n+1)!}{(n-2)!}$ (c) $\dfrac{(n+6)!}{(n+2)!}$ (d) $\dfrac{(n+1)!(n+2)!}{(n!)^2}$
7. (a) 8(6!) (b) 1443(7!) (c) 89(8!) (d) 116(3!)
8. (a) $(n-1)(n-1)!$ (b) $(n^2-n-1)(n-2)!$ (c) $(n^2+n-1)(n-1)!$ (d) $(n^2+4n+4)n!$
9. $\dfrac{d}{dx}(\ln kx)=\dfrac{d}{dx}(\ln k+\ln x)=0+\dfrac{1}{x}=\dfrac{1}{x}$
10. $\dfrac{n}{mx}$
11. (a) $2x\ln x+x$ (b) $x^2e^{2x}(3+2x)$ (c) $e^{-4x}(1-4x)$ (d) $e^{2x}+2xe^{2x}+2x\ln x+x$
12. (a) $\sin x+x\cos x$ (b) $2x\cos x-x^2\sin x$ (c) $e^x\cos(3x+1)-3e^x\sin(3x+1)$
    (d) $e^{-2x}\cos(4x-3)-2\,e^{-2x}\sin(4x-3)$
13. (a) -20 (b) 24.2 (c) $\dfrac{1}{3}(x^2+2)^{\frac{3}{2}}+C$ (d) 539
14. (a) $\dfrac{2}{3}(e^x+2)^{\frac{3}{2}}+C$ (b) 78 (c) $\dfrac{(\ln x)^2}{2}+C$ (d) 0.461
15. (a) $\ln(x-5)+C$ (b) 0.373 (c) $\dfrac{5}{2}\ln(2x+1)+C$ (d) 0.678
16. (a) -0.442 (b) $-\cos(x-5)+C$ (c) -0.001 (d) -0.060
17. 385
18. 1.198 862
19. 797 148
20. $\displaystyle\sum_{n=1}^{16}(7n-5),\ 872$

Answers to Problems

CHAPTER 1

1. -787 kJ mol^{-1}

2. $V = -\dfrac{Ze^2}{4\pi\varepsilon_o r}$

3. $T = \dfrac{\Delta H - \Delta G}{\Delta S}$

4. (a) $K = \dfrac{[CH_4]}{[CO][H_2O]^2}$

   (b) $K = \dfrac{[CH_4][H_2O]}{8[CO]^4}$

   (c) $K = \dfrac{8[CH_4][H_2O]}{[CO]^4}$

   (d) $K = \dfrac{[CH_4][H_2O]}{[CO]^4}$

5. (a) $kt = \dfrac{1}{[NO_2]} - \dfrac{1}{[NO_2]_o}$

   (b) $kt = \dfrac{[NO_2]_o - [NO_2]}{[NO_2][NO_2]_o}$

6. $\dfrac{1}{\phi} = \dfrac{k_F + k_{IC} + k_{ISC}(k_R[R] + k_P + k'_{ISC})}{k_{ISC}k_R[R]}$

7. $\mu = \left(\dfrac{\sigma^2}{\pi}\right)\left(\dfrac{8kT}{z_{AB}^2}\right) N_A^4 [A]^2 [B]^2$

8. (a) $c = 2.998\,000\,00 \times 10^8 \text{ m s}^{-1}$
   (b) $R = 1.097\,000\,0 \times 10^7 \text{ m}^{-1}$
   (c) $F = 9.649\,0 \times 10^4 \text{ C mol}^{-1}$
   (d) $p = 1.013\,25 \times 10^5 \text{ Pa}$
   (e) $V_o = 2.241\,41 \times 10^{-2} \text{ dm}^3$

9. (a) $V = 2.196 \times 10^3 \text{ m}^3$
   (b) $p = 1.86 \times 10^4 \text{ Pa}$
   (c) $E = 3.142 \times 10^3 \text{ V}$
   (d) $k = 3.84 \times 10^5 \text{ dm}^3 \text{ mol}^{-1} \text{ s}^{-1}$
   (e) $n = 8.544 \times 10^{-5} \text{ mol}$

10. (a) $\tilde{v} = \dfrac{3RZ^2}{4}$

   (b) $\tilde{v} = \dfrac{5RZ^2}{36}$

   (c) $\tilde{v} = \dfrac{3RZ^2}{4n_1^2}$

11. $10.33 \text{ J K}^{-1}\text{mol}^{-1}$

12. (a) $v = k[O_3]^{\frac{4}{3}}$

   (b) $v = \dfrac{k[O_3]^{\frac{4}{3}}}{2^{\frac{2}{3}}}$

   (c) $v = 0.072\, k\,\text{mol}^{\frac{4}{3}}\,\text{dm}^{-4}$

CHAPTER 2

1. (a) (i) 1.54 Å (ii) 1.54 Å
   (b) (i) 25.0°C (ii) 25.01°C
   (c) (i) −433 kJ mol^{-1} (ii) −432.88 kJ mol^{-1}
   (d) (i) 16.0 (ii) 16.00
   (e) (i) 8.31 J K^{-1}mol^{-1} (ii) 8.31 J K^{-1}mol^{-1}
2. (a) 2.8 g cm^{-3} (b) 15.03 g (c) 159 kJ mol^{-1}
   (d) 7.9 kJ (e) 0.042 mol dm^{-3}
3. (a) −283 kJ mol^{-1} (b) 35.21 J K^{-1}mol^{-1} (c) 133 dm^3 mol^{-1}
4. 1.001 2 or 1.001 19
5. ±0.2 J K^{-1}mol^{-1}, ±0.1 J K^{-1}mol^{-1}
6. 5.64 ± 0.04 m^3
7. 3.07 ± 0.05%
8. 589.592 4, 589.594 3 nm
9. mean = median = 494 kJ mol^{-1}, mode = 493 kJ mol^{-1},
   variance = 4 (kJ mol^{-1})^2, standard deviation = 2 kJ mol^{-1}
10. 0.007 Å
11. 0.001 dm^3
12. 0.7 kJ mol^{-1}

CHAPTER 3

1. (a) $T = \dfrac{\Delta G - \Delta H}{\Delta S}$ (b) $V_1 = \dfrac{V - n_2V_2}{n_1}$ (c) $P = C - F + 2$ (d) $p_1 = \dfrac{p_2V_2}{V_1}$

2. (a) $\dfrac{nR}{V}$ (b) $\dfrac{RT}{V}$ (c) $\dfrac{RT}{p}$

3. Plot C_p against T: gradient = b, intercept = a

4. $\dfrac{\Delta H^{\ominus}}{RT^2}$

5. $\dfrac{dCp}{dT} = b - \dfrac{2c}{T^3}$

6. 411 m s^{-1}

7. $\left(\dfrac{nR}{p}\right)dT - \left(\dfrac{nRT}{p^2}\right)dp$

8. $a = 0.138$ Pa m^6 mol^{-2}, $b = 3.18 \times 10^{-5}$ m^3 mol^{-1}

9. $pV = RT$

10. $\alpha = 0.9$

11. $V = \dfrac{RT \pm \sqrt{R^2T^2 - 4ap}}{2p}$

12. $x = 0.176$

13. Units on each side are J.

14. Units on each side are K.

CHAPTER 4

1. (a) 0.063 mol (b) 4.00 dm^3

2. 31.8 g

3. 9.4 cm^3

4. 0.269

5. (a) 0.080 (b) 0.8

6. 160 mm Hg

7. 0.18

8. (a) π against c, gradient $= RT$, intercept $= 0$

   (b) log k against $\sqrt{I}$, gradient $= 1.02\, z_A\, z_B$, intercept $= \log k_o$

   (c) Λ against $\sqrt{c}$, gradient $= -(P + Q\Lambda_o)$, intercept $= \Lambda_o$

9. (a) Plot E against ln Q, gradient $= -\dfrac{RT}{zF}$, intercept $= E^{\ominus}$

   (b) Plot E against T, gradient $= -\dfrac{R}{zF}\ln Q$, intercept $= E^{\ominus}$

10. Gradient $= \dfrac{1}{K\Lambda_o^2}$, intercept $= \dfrac{1}{\Lambda_o}$

11. $\dfrac{\partial c}{\partial t} = D\dfrac{\partial^2 c}{\partial x^2}$

12. (a) $V_1 = \dfrac{1}{\rho}$ (b) $\Delta T_b = \dfrac{K_b m}{2}$ (c) $\alpha = 1$, complete dissociation

13. 17.6 cm^3 mol^{-1}

14. 1 005.65 cm^3

15. Minimum at $x = 0.433$.

CHAPTER 5

1. $4.6 \times 10^{-10}\,\text{mol dm}^{-3}\,\text{s}^{-1}$

2. $\dfrac{d[NH_4Cl]}{dt} = \dfrac{1}{3}\dfrac{d[Cl_2]}{dt} = -\dfrac{d[NCl_3]}{dt} = -\dfrac{1}{4}\dfrac{d[HCl]}{dt}$

3. $\dfrac{1}{1-a}\left([A]^{1-a} - [A]_o^{1-a}\right) = kt$, not valid when $a = 1$.

4. $\log v = \log k + m \log [NO] + n \log [Cl_2]$

   Keep the concentration of one reactant constant. Plot $\log v$ against $\log c$ where c = concentration of other reactant. Order (m or n) is given by gradient.

5. First order, $k = 6.61 \times 10^{-6}\,\text{s}^{-1}$.

6. $t_{\frac{1}{2}} = \dfrac{1}{k[A]_o}$

7. 13 hours

8. $6.3 \times 10^3\,\text{s}^{-1}$

9. $81\,\text{kJ mol}^{-1}$

10. $-190\,\text{kJ mol}^{-1}$

11. $[Br] = \sqrt{\dfrac{k_1}{k_5}[Br_2]}$

12. $[NO] = \dfrac{k_2[NO_2]}{k_3}$

CHAPTER 6

1. $\dfrac{\pi}{6}$

2. $1.47\,\text{Å}$

3. $1.922\,\text{Å}$

4. $0.145\,\text{nm}$

5. $3.150\,\text{Å}$

6. $36.5°$

7. $110.2°$

8. $a\mathbf{i}$, $a\mathbf{j}$, $a\mathbf{k}$, $\frac{1}{2}a\mathbf{i} + \frac{1}{2}a\mathbf{j} + \frac{1}{2}a\mathbf{k}$

9. $364.3\,\text{Å}^3$

10. (a) $F(0\,0\,l) = f_1 \exp(2\pi ilz) + f_2 \exp(2\pi il(z + 0.5))$
    (b) $\cos 2n\pi = 1$ so $F(00l) \neq 0$ when $l = 2n$.

11. B $(1.2, -0.7)$, C $(0.0, -1.4)$, D $(-1.2, -0.7)$, E $(-1.2, 0.7)$, F $(0.0, 1.4)$

12. $\begin{pmatrix} x_2 \\ y_2 \end{pmatrix} = \begin{pmatrix} -1 & 0 \\ 0 & 1 \end{pmatrix}\begin{pmatrix} x_1 \\ y_1 \end{pmatrix}$

CHAPTER 7

1. $5.291\,764\,0 \times 10^{-11}\,\text{m}$

2. Minimum at $r = \sqrt[3]{\dfrac{3A}{2B}}$

3. $arc\ E(n) = \dfrac{2d}{h}\sqrt{mn}$

4. $v - v_o = 2.52 \times 10^{14}\ T_{\max}$ where $v_o = 5.47 \times 10^{14}\ Hz$

5. $E_1 = \dfrac{1}{2}hv_o,\ E_2 = \dfrac{3}{2}hv_o,\ E_3 = \dfrac{5}{2}hv_o\ E_4 = \dfrac{7}{2}hv_o\ E_5 = \dfrac{9}{2}hv_o$

6. 80 J mol^{-1}

7. $arc\ \Psi_{1s}(r) = -\left(\dfrac{a_o}{Z}\right)\ln\left[\dfrac{r}{2}\left(\dfrac{a_o}{Z}\right)^{\frac{3}{2}}\right]$

8. $\dfrac{dm}{dE} = \dfrac{\pi}{h}\left(\dfrac{2I}{E}\right)^{\frac{1}{2}}$

9. $A = \dfrac{1}{\sqrt{2\pi}}$

10. $\dfrac{a^3 h^2}{24\pi^2 m}$

11. $\dfrac{a^2}{3} - \dfrac{a^2}{2n^2\pi^2}$

12. $\dfrac{3}{2}a_o$

13. $E = \alpha + \beta,\ E = \alpha - \beta$

14. $E = \alpha + 1.62\ \beta,\ E = \alpha - 1.62\ \beta,\ E = \alpha + 0.62\ \beta,\ E = \alpha - 0.62\ \beta$

CHAPTER 8

1. 4.23 D, -1.74×10^{-39} C m^2

2. $0.775e$

3. 6.949×10^{13} Hz, 4.314×10^{-6} m, 4.605×10^{-20} J

4. $I = I_o e^{-bl}$

5. 3.1×10^{-5} mol dm^{-3}

6. (a) $F(J + 1) - F(J) = 2B\ (J + 1) - 4D(J + 1)^3$

    (b) $8.417\ 27 \times 10^{-22}$ J

7. 6.303 cm^{-1}

8. (a) $G(1) - G(0) = \omega_e - 2\ \omega_e\ x_e$

    (b) $3.452\ 1 \times 10^{-6}$ m

9. $3.796 - 0.035$ m

    For positive values of m, a value will be subtracted from 3.796 so the value of $\tilde{v}\ (m + 1) - \tilde{v}\ (m)$ decreases. For negative values of m, a value will be subtracted from 3.796 so the value of $\tilde{v}\ (m + 1) - \tilde{v}\ (m)$ increases.

10. $B\ (4J + 6)$

11. The peak due to $(CH_3)_2$ will be split into two in the ratio 1:1. The peak due to CHI will be split into seven in the ratio 1:6:15:20:15:6:1.

12. $g(\omega) = \dfrac{1}{2\pi(a + i\omega)}\left[1 - e^{-(a+i\omega)}\right]$

CHAPTER 9

1. 30
2. (a) $N!$ (b) N
3. $p = \dfrac{NkT}{V}$
4. 9.79×10^{29}
5. (a) $\ln N! = N \ln N - N$ (b) $N^N e^{-N}$ (c) 14%
6. $q' = q \exp\left(\dfrac{\alpha}{kT}\right)$
7. 4
8. (a) $2.104\ 38 \times 10^{-22}$ J (b) 19.561
9. 685×10^{34}
10. (a) 5.409×10^{-7} (b) 9.831×10^{-6}
11. $7.562 \times 10{-22}$ J K^{-1}
12. 4.496×10^{3}

Chemical Index

absolute temperature *see* temperature, absolute
absorbance, measurement of 27, 28, 296–298, 318
absorption coefficient 296, 318
absorption of radiation 296–319
acetaldehyde, catalytic oxidation of 110
acetic acid *see* ethanoic acid
acetic anhydride
    production 110
    reaction with methanol 174
acid strength 110
activation energy 67, 186–190, 200
activity 109, 126
activity coefficient 109, 111, 126, 131, 133, 146
aldehyde 51
alkene 205–206
allyl radical 280–281
air
    composition of 130
    partial pressure of 146
    as a source of nitrogen 113
ammonia
    decomposition of 156–157
    formation of 113
amount of substance 12, 34, 41, 56, 64, 70, 90, 91, 93–96, 123, 124, 125, 127, 128, 137, 145, 148, 150
Ångström, unit 49, 92, 343
Arrhenius equation 185–192, 200
asymmetric top 298–299
atmosphere, unit 2, 30, 31, 84–88, 92–93, 113, 120
atomic mass 16, 33, 124
atomic mass, relative 55, 291–292
atomic scattering factor *see* scattering factor, atomic
Avogadro constant 92, 302, 326, 345
Avogadro's law 96
axis of rotation 228, 231–232, 236–237, 298, 331
azimuthal angle 263
azomethane, decomposition of 199

Balmer series 262
barometer 131
Beattie–Bridgeman equation 93–94
Beer–Lambert law 296–298
beryllium hydride, dipole moment of 289
blast furnace 33–34
body-centred cubic lattice 201–204, 236
boiling point 84–85, 86
    and elevation of 121, 127, 147
Boltzmann constant 24, 92, 200, 327, 334, 341, 342, 345
Boltzmann distribution law 323, 325
Boltzmann equation 325–329
Bohr radius 284, 285–286, 345

bond
    aluminium-hydrogen 319
    angle 219–222, 228, 235–236, 285
    barium-nitrogen 207
    carbon-carbon 206
    carbon-chlorine 59–60
    carbon–hydrogen 59–60, 205–206, 231–232, 289–290
    chlorine-chlorine 59–60
    copper–nitrogen 56
    copper–oxygen 235
    enthalpy 58–60
    equilibrium length 244–246
    force constant 244–246, 301
    hydrogen–chlorine 49–50, 59, 318
    hydrogen–fluorine 291–292
    length 49–50, 55
    potassium-bromine 318
Born–Haber cycle 23, 60–63
boundary condition 251, 260
Boyle's law 96
Bragg's law 207–213, 223
bromine, reaction with hydrogen 152–153, 200
buffer solution 151
burette 28, 32, 37, 39–40
butadiene
    dimerization of 170
    Hückel determinant of 287
    structure of 287

$C=O$ stretching frequency 51–52
calcium, crystal structure 202
calorie, unit 92
calorimeter 30
carbon dioxide
    formation of 19, 33, 40
    root-mean-square speed of 119
    sublimation of 75
carbon monoxide
    in preparation of propionaldehyde 151
    reaction with iron (III) oxide 33
    rotation-vibration spectrum of 319
    van der Waals constants of 93
carbonyl compound 51, 173
carboxylate ion 51
carboxylic acid 51
celsius scale *see* temperature, celsius
cell *see* electrochemical cell
centre of mass 291–292
centre of symmetry 230
charge, atomic *see* electronic charge

Mathematical Index

Printed in the United States
by Baker & Taylor Publisher Services